T0146090

Genocide and the
Geographical Imagination

Genocide and the Geographical Imagination

Life and Death in Germany, China, and Cambodia

James A. Tyner

ROWMAN & LITTLEFIELD PUBLISHERS, INC.
Lanham • Boulder • New York • Toronto • Plymouth, UK

Published by Rowman & Littlefield Publishers, Inc.
A wholly owned subsidiary of The Rowman & Littlefield Publishing Group, Inc.
4501 Forbes Boulevard, Suite 200, Lanham, Maryland 20706
www.rowman.com

10 Thornbury Road, Plymouth PL6 7PP, United Kingdom

British Library Cataloguing in Publication Information Available

Library of Congress Cataloging-in-Publication Data
Tyner, James A., 1966–
  Genocide and the geographical imagination : life and death in Germany, China, and
Cambodia / James A. Tyner.
    p.   cm.
  Includes bibliographical references and index.
  ISBN 978-1-4422-0898-8 (alk. paper) — ISBN 978-1-4422-0900-8 (electronic)
  1. Genocide—Germany—History. 2. Genocide—China—History. 3. Genocide—
Cambodia—History. I. Title.
  D804.G3T96 2012
  364.15'1—dc23
                                                                        2012012419

∞™ The paper used in this publication meets the minimum requirements of American
National Standard for Information Sciences—Permanence of Paper for Printed Library
Materials, ANSI/NISO Z39.48-1992.

Printed in the United States of America

Contents

List of Illustrations   vii

Preface and Acknowledgments   ix

1   The Spatiality of Life and Death   1

2   The State Must Own Death: Germany   33

3   Starving for the State: China   81

4   Normalizing the State: Cambodia   111

5   Everyday Death and the State   153

Selected Bibliography   165

Index   177

About the Author   181

Illustrations

1.1	Hartheim Castle, Austria	2
1.2	A member of an *Einsatzgruppe* prepares to shoot a Ukrainian Jew kneeling on the edge of a mass grave	9
2.1	A Jewish woman walks toward the gas chambers with three young children at Auschwitz-Birkenau	34
2.2	Jewish women and children selected for the gas chambers at Auschwitz-Birkenau	34
2.3	Eugenics poster	49
2.4	Propaganda slide featuring a disabled infant	50
2.5	Propaganda slide featuring two doctors working at an asylum for the mentally ill	54
2.6	Twenty-three-year-old Elizabeth Killiam, the mother of twins, was sterilized in Weilberg	57
2.7	The corpse of a woman put to death as part of the Operation T-4 euthanasia program	64
2.8	Photograph of a survivor at the Hadamar Institute judged insane for having a Jewish boyfriend	65
2.9	Mass execution and grave site associated with the liquidation of the Mizocz ghetto, Poland	68
2.10	Corpses of prisoners exhumed from a mass grave in the vicinity of Hirzenhain, Germany	69
2.11	Newly arrived prisoners, with shaven heads, stand at attention during a roll call in the Buchenwald concentration camp	70
2.12	Corpses at the recently liberated Dachau concentration camp	72
3.1	Propaganda photograph of peasants harvesting rice in the province of Guangdong during the Great Leap Forward	91

3.2   Propaganda photograph of women harvesting cocoons from
       silkworms during the Great Leap Forward   92
3.3   Propaganda photograph of women operating an irrigation
       system during the Great Leap Forward   96
4.1   Staff of Tuol Sleng Security Center dining with their fami-
       lies   117
4.2   Two female Khmer Rouge cadre stand at attention near a
       work camp   118
4.3   Transporting dirt on an irrigation project   121
4.4   Women working on an irrigation project   124
4.5   Men working on an irrigation project along the Chinith
       River, Kampong Thom Province   125
4.6   Construction on an irrigation project   126
4.7   Unidentified prisoner at S-21   128
4.8   Unidentified prisoner who committed suicide at S-21   128
4.9   Mug shot of Hout Bophana, arrested on October 10, 1976   129
4.10   Young girls preparing "rabbit dropping" medicine   136
4.11   Young girls preparing "rabbit dropping" medicine   137
4.12   Mass grave site at one of Cambodia's many killing fields   139

Preface and Acknowledgments

Tuol Sleng prison, located in Phnom Penh, Cambodia, was one of many security centers erected by the Khmer Rouge throughout its genocidal reign. At this site, designated "S-21," prisoners were photographed, detained, interrogated, tortured, and executed. Approximately 13,000 men, women, and children entered Tuol Sleng; fewer than 200 are known to have survived.

Keat Sophal, the woman whose image is on the cover of this book, was one of the victims. We know little about her death, and even less about her life. Documentary evidence indicates that she was arrested on April 13, 1977. She was detained at S-21 for ninety-nine days until the day of her "termination" on July 22.

We know that she was a Khmer Rouge cadre; her job was to take care of children. No information has been forthcoming about her family or what she did prior to the time of the genocide. We do not know why she was arrested or killed. In all likelihood, she was interrogated and tortured; perhaps she was raped. No confession or record of her "crimes" remain. Was she found "guilty" of traitorous activities to the state? Was she found "delinquent" in her patriotic duties? Or was Sophal simply arrested because she was associated with someone else charged of a crime? It is not known. It is also not known whether she died at S-21, or was taken to the nearby killing fields, Choeung Ek, to be killed.

Two dates—a date of arrest and a date of termination—and a photograph. This is all that remains of the life and death of Keat Sophal. Her facial expression suggests resignation. As a Khmer Rouge cadre, Sophal most likely knew of her eventual fate. By the summer of 1977, she had probably witnessed many deaths and was well aware that once accused by Angkar her fate was sealed. In Cambodia, during the time of the genocide, to be accused was to be guilty; to be guilty was to be sentenced to death.

Keat Sophal was one of approximately two million people who died throughout the Cambodian genocide. And in certain respects, her legacy is a microcosm of that event—and of this book. Genocides are not inevitable; nor are these "natural" events that take place. Rather, genocides result from the constellation of myriad decisions: calculations of life and death, of who may live and who will be disallowed life to the point of death. Someone, at some point, determined that Keat Sophal was required to die, that her death was more important to the state than was her life. And because of that decision, she was executed.

Genocides are so named because they entail the killing of aggregates; the murder of masses of people. As such, when we speak of genocide, we tend to invoke large numbers. These biostatistics, however, belie the inhumanity of genocide, for they too often hide the fact that the edifice of genocide is built on the bodies—literally the skeletons—of individuals, individuals like Keat Sophal. It is ultimately that relationship between the state and the individual that comprises genocide. This relationship establishes the calculus—the bio-arithmetic—by which a person's life is determined and by which a sovereign state makes a decision to let live, to kill, or to deny life to the point of death. And that is what this book is about. States do not embark on a course of genocide per se; rather, they embark on policies and practices that relate directly, intimately, to the possibility of life or death for a person.

Genocide and the Geographical Imagination is about life and death; but it is also about space. In the chapters that follow I argue that those events labeled genocide are predicated upon a violent enforcement of the will to space. Mass violence, I maintain, results from the imposition of state-sanctioned normative geographical imaginations that justify and legitimate unequal access to life and death. A geographical perspective on genocide highlights that mass violence, in the imaginations of perpetrators, is viewed as an effective—and legitimate—strategy of state-building.

The genesis of this book dates to the summer of 2001—my first venture to Cambodia. Over the next decade, I have made repeated journeys back to the country; and each visit brings with it new experiences, new insights. What remains constant, however, is the dark legacy of genocide. It is exceptionally humbling to witness the changes that have taken place in Cambodia, knowing well the violence that engulfed this region for much of the latter twentieth century. It is hoped that, in some small way, this book might provide some sense of understanding of the seemingly senseless violence that tore apart Cambodia. Accordingly, immediate thanks are extended to Susan McEachern and Grace Baumgartner at Rowman & Littlefield, for their guidance, assistance, and overall support of this project.

Over the years I have been fortunate to work with, and share ideas with a number of exceptional scholars; some were my advisors, others were my students, and still others have been colleagues. All have been important. At the Documentation Center of Cambodia, I am particularly grateful for the assistance and insights of Youk Chhang, Khamboly Dy, Socheat Nhean, and Kok-Thay Eng. Special thanks are extended to my translators, Kuy Bunrong, Y Manoka, Keo Ratanatepy, and Visal Veng. At the United States Holocaust Memorial Museum, I thank Caroline Waddell and Dave Bair. At Kent State University, a special note of gratitude is extended to my chair, Mandy Munro-Stasiuk, for her support—both financially and in spirit—of my research. Thanks also to the many individuals who read, in whole or in parts, earlier versions of the manuscript: Gabriela Brindis Alvarez, Bradley Austin, Alex Colucci, Stephen Egbert, Colin Flint, Joshua Inwood, Alex Peimer, Stian Rice, Dave Stasiuk, and Dave Widner.

Closer to home, literally, I thank my parents, Dr. Gerald Tyner and Dr. Judith Tyner, for their continued support; thanks also to my brother, David Tyner, and my aunt, Karen (Tyner) Owens. All of these remarkable individuals, in their own way, helped provide a solid moral foundation for my life and my work. As always, I thank my ten-year-old puppy, Bond, and my twelve-year-old cat, Jamaica, for their late-night support of my writing. Most important, however, I thank those who have endured long periods of absence (and periods of absentmindedness) while writing this book: Belinda, my friend, partner, and wife; and Jessica and Anica Lyn, our two daughters. Throughout the course of this project, they have provided a balance: to see the simple joys in life that are too often taken for granted in our day-to-day lives. And it is with their blessings that I dedicate this book to the memory of Keat Sophal and to those who suffered and died needlessly in Germany, China, and Cambodia.

Chapter One

The Spatiality of Life and Death

Located in the town of Alkoven, Austria, sits Hartheim Castle. The tranquil silence surrounding this stately structure—built over three centuries ago—belies its recent, brutal, past. Between May 1940 and December 1944 the castle was used as a "killing center" by the Nazi state as part of a broader euthanasia program designed to "eliminate" mentally and physically handicapped adults. At the end of World War II, documents presenting detailed financial calculations of the "savings" achieved through the euthanasia program were found in a safe in the castle. Nazi officials had estimated that the "disinfection" of 70,273 individuals resulted in a savings of over 245,955 reichsmarks per day. More chilling, perhaps, is that these calculations were based on the savings of food items. In other words, the murder of these people contributed to such savings as 239,067 kilograms of marmalade and 653,516 kilograms of meats and sausages.[1] To be blunt, the lives of over 70,000 people were measured in jellies and frankfurters.

To calculate the value of human life by the equivalent foodstuffs is the epitome of a dehumanization of life. However, it is much, much more. Indeed, such an arithmetic rationale for murder speaks directly to the valuation of life. Note that we are not referring to the "meaning" of life per se. We are not (yet) questioning whether any particular life is worth living. Rather, we are simply confronted with a cost-benefit analysis, where the taking of a life results in the direct savings of other commodities. Once it has been determined that certain financial or material resources might be saved through the elimination of people—then it becomes necessary to determine which lives are worthy or not.

We read the above calculated management of life with horror—in part because we know the ending of the story. The involuntary sterilization and euthanasia programs conducted in Nazi Germany during the 1930s led directly

1

Figure 1.1.   Hartheim Castle, Austria. One of six hospitals used by the Nazis as part of their "euthanasia" program. *Source:* United States Holocaust Memorial Museum

(though not inevitably) to the Holocaust of the 1940s. And equipped with this knowledge, we respond to the trade-off between groceries and human life as an *evil*, God-less act. Indeed, it was Primo Levi, a Holocaust survivor, who wrote: "There was Auschwitz, therefore God cannot exist."

But these monetary renderings are not the exclusive domain of the Nazi Party. Nor are they limited to genocide. They are, in fact, a product of the modern bureaucratic state. Consider, for example, that following the devastating attack on the World Trade Center and the Pentagon on September 11, 2001, in the United States, families of victims were allocated cash payouts. Within six months, by March 2002, the US government issued guidelines for compensating the next of kin for victims of the terrorist attacks. As David Dranove explains, the government divided compensation into two parts—an economic component based on the victim's income-earning potential, and a noneconomic component. The monetary value of the economic component averaged US$1.6 million, while the noneconomic component was set at US$250,000.[2]

And this calculated management and valuation of life need not occur only in times of national tragedy. Such decisions are exceptionally banal and happen *every day* in many countries around the world. Again, turning to Dranove, we find that routinely, ordinary people in the United States and elsewhere put a dollar value on life and limb; they are jurors in civil lawsuits in which an individual or firm is accused of negligently causing injury or death. And if a defendant is found negligent, the jury must calculate a cash award to compensate the victim or next of kin.[3]

In the context of the Holocaust, the calculation of life appears as pure evil. But how do we square this with other forms of monetary payout? Why is one calculation considered immoral and godless, while the other is considered acceptable? What does this say more broadly about our own calculations (and valuation) of life and death? What can genocide and mass violence say about how people are valued?

In *Genocide and the Geographical Imagination* I look back at genocide with an eye to the present. This is not a nostalgic look—to long for something so horrific is morally discordant. So why look? For many authors, the past reveals a path to the future. Following this reasoning, we hope that an understanding of previous episodes of mass violence might provide guidance in the prevention of future violence. Such a goal is found, for example, in the writings of David Hamburg.[4] Hamburg maintains that "all the genocides of the twentieth century were clearly visible years in advance, but largely dismissed, even denied, by the international community until mass killing was well under way." He writes with optimism, noting that the "paths of genocide-prone behavior are clear; we are learning how to provide help and apply pressure at strategic points along those paths to prevent it."[5] Ben Kiernan concurs, explaining that "six hundred years of evidence helps us detect [genocide's] essential elements not only in retrospect but, by analysis of common causes, potentially in advance, which increases the possibility of preventing future genocides with timely action."[6]

For other authors, the past provides lessons—all sorts of lessons—for our understanding of various concepts. Zygmunt Bauman finds answers to modernity through his study of the Holocaust; Martin Shaw derives insight into war through genocide.[7] While sympathetic to these efforts, I remain somewhat guarded as to the lessons we might learn from mass violence. Education is by definition a future-oriented and optimistic activity.[8] But there is a danger that we fall back on platitudes; that we conclude that some "good" can be found in "evil." And we also forget that education was used to justify mass violence. During the Cambodian genocide, for example, school lessons might consist of public torture and execution. Roeun Sam was a fourteen-year-old girl when her nightmare began. She describes how the Khmer Rouge assembled a group of children within a temple at a place called Thunder Hill. While the children sat obediently, silently, two adult prisoners were brought forward. A Khmer Rouge official informed the children that "if someone betrays Angka,[9] they will be executed. We want everyone to know that these people are bad examples, and we don't want other people to be like this." Sam explains that the Khmer Rouge forced the children to sit in front of the two prisoners; the children were warned, "If anyone cries or shows empathy or compassion for this person, they will be punished by receiving the same treatment."[10] Sam describes what happened next:

> Angka told someone to get the prisoner on his knees. The prisoner had to confess what he had done wrong. Then the prisoner began to talk but he didn't confess anything. Instead, he screamed, "God, I did not do anything wrong. Why are they doing this to me? I work day and night, never complain, and even though I get sick and I have a hard time getting around, I satisfy you so you won't kill people. I never thought to betray Angka. This is injustice. . . ." Suddenly one of the [Khmer Rouge guards] hit him from the back, pushed him, and he fell face to the ground. It was raining. We sat in the rain, and then the rain became blood. He was hit with a shovel and then he went unconscious and began to have a seizure. Then Angka took out a sharp knife and cut the man from his breastbone all the way down to his stomach. They took out his organs. . . . They tied the organs with wire on the handlebars of a bicycle and biked away, leaving a bloody trail.[11]

It is difficult to know what lesson Sam learned from this experience; and, looking back, it is difficult to know what lesson we might take away. Where are we to find meaning in such cruelty? Are we to even find meaning? At this point, I am also circumspect—and perhaps not so optimistic. Too often we either attempt to explain past genocides to predict the next, or we explain away genocide as simply evil. For me, neither is satisfactory. Here, I propose a different tack. Inverting the question, I wonder what genocide might say about *sovereignty and the spatiality of life and death.*

Humans are social beings; but humans are also spatial beings. This is not meant to be a glib statement that we exist in space, but rather a recognition that through our ordinary daily activities—of going to work, going to school, shopping for groceries—we encounter other people and other "spaces." Our thoughts, emotions, and indeed our actions are influenced by these encounters. At the same time, however, our own actions reflect upon those encounters. Just as we are transformed through our daily activities, so too are the "spaces" that we inhabit transformed. In other words, we *produce* and are *produced* by space. However, as Henri Lefebvre cautions, space is not produced in the sense that a kilogram of sugar or a yard of cloth is produced.[12] Space, as such, has no materiality and exists conceptually as a result of human relations and interactions. As Cresswell explains, the "social and the spatial are so thoroughly imbued with each other's presence that their analytic separation quickly becomes a misleading exercise."[13]

However, we do not always have adequate words to think through these sociospatial relations. Consequently, Ed Soja has introduced the term "spatiality" in reference to socially produced space. Soja explains:

> The dominance of a physicalist view of space has so permeated the analysis of human spatiality that it tends to distort our vocabulary. Thus, while such adjectives as "social," "political," "economic," and even "historical" generally suggest . . . a link to human action and motivation, the term "spatial" typically evokes a physical or geometrical image, something external to the social context and to social action, a part of the "environment," a part of the setting for society—its naively given container—rather than a formative structure created by society. We really do not have a widely used and accepted expression in English to convey the inherently social quality of organized space, especially since the terms "social space" and "human geography" have become so murky with multiple and often incompatible meanings.[14]

An emphasis on spatiality—the purposefully organized space of social interactions—opens the possibility to work through the intimate relationships between and imaginings of life, death, and place. Over six decades ago John Wright wrote of terra incognitae: places that are geographically unknown.[15] Terra incognitae are both literal and symbolic: they constitute those shadowy spaces just beyond our reach, beckoning us as do sirens, drawing us ever farther from shores of comfort. For Wright, "Voyages into this shadow became a favorite theme of poets and story tellers—the theme of the Argonautic myth and the *Odyssey*, of the legends of Sinbad and Saint Brandan."[16] The theme also, we might add, of Dante as he was led by Virgil through the nine circles of Hell before ascending, ultimately, to the mountain of lightness. For me, genocide is geographically unknown; it is terra incognita, but not in the sense of an unexplored, or unvisited place. Indeed, human history reveals that genocide has been visited far too often. Rather, genocide is terra incognita in the sense that the geographical imaginations of life and death—on a massive scale—remain opaque.

The geographical imagination. According to Edward Said, geographic imaginations are "ways of seeing" that "legitimate a vocabulary, a representative discourse peculiar to the understanding of places that becomes the way in which a place is known."[17] Imaginative geographies are, in effect, "metageographies" that provide the foundation for sociospatial actions: the spatiality of everyday life. As explained by Lewis and Wigen, these may be conceived as spatial structures through which people order their knowledge of the world. Whether one considers diplomats, politicians, and military strategists—or the peasants, factory workers, and farmworkers—all employ particular geographical imaginations in their pursuit of goals.[18] And through these imaginations, space becomes politicized. Derek Gregory is clear on this point: geographical imaginations involve a politics of space. He asserts, "Who claims the power to represent: to imagine geography like this rather than like that? The process of articulation is . . . also a process of valorization."[19] In the following chapters I provide three spatial stories—not histories, in the conventional sense of the term—but instead deliberate geographically informed readings of Germany, China, and Cambodia. This is not, to be clear, an introductory text to genocide, nor a chronology of these specific genocides. As stated earlier, my intent is to work through genocide as a means of providing insight into the larger questions of sovereignty and the spatiality of life and death. In the remainder of this chapter I provide a more solid conceptual foundation for this task.

VIOLENCE AND THE WILL TO SPACE

Humans are apparently unique in their ability to kill members of their own species—often on a scale that borders on the unimaginable.[20] Throughout the twentieth century approximately 230 million people died in wars and other forms of mass conflict. During World War I, for example, an estimated 13 to 15 million people died because of political decisions that led Germany, France, Russia, Austria-Hungary, and other states into war. The Armenian genocide, it should be remembered, is included in this conflict. World War II, likewise, contributed to the death of between 65 and 75 million people. Embedded within this conflict are the estimated 6 million Jews who perished in the Holocaust.[21]

And the twentieth century is far from unique in its prevalence of mass violence. Lars Svendsen notes that over the last 3,400 years, there were only 243 years, altogether, in which humankind was *not* at war.[22]

What accounts for humans' ability to engage in such large-scale violence? What allows (or impels) humans to kill one another? There are many existent models, theories, and frameworks that seek to account for this violence. Notable are the works of Leo Kuper, Ervin Staub, Martin Shaw, and Alexander Hinton. What appears to hold constant is recognition that killing—ranging from homicide to genocide—is not an irrational act *from the standpoint of the perpetrator*. As James Gilligan concludes, "even the most apparently 'insane' violence has a rational meaning to the person who commits it."[23]

The killing of humans by other humans is neither natural nor inherent. Although genetic evolution may have contributed to a propensity to engage in violence, including killing, this does not mean that humans are natural-born killers. As Daniel Chirot and Clark McCauley write, "all but those most habituated to extreme brutality or a small number who seem to lack normal emotional reactions to bloody violence, have to overcome a sense of horror when they engage in or witness slaughter firsthand."[24] In fact, numerous studies on the psychology of combat-related killing—including studies of Nazi atrocities committed during the Holocaust—have demonstrated that humans are exceptionally reticent to kill.[25] Dave Grossman, for example, finds that "there is ample evidence of the resistance to kill and that it appears to have existed at least since the black-powder era. This lack of enthusiasm for killing the enemy causes many soldiers to posture, submit, or flee, rather than fight; it represents a powerful psychological force on the battlefield; and it is a force that is discernible throughout the history [of warfare]."[26]

Soldiers—and people in general—do not readily kill; why not?[27] According to Grossman, "Looking another human being in the eye, making an independent decision to kill him, and watching as he dies due to your action combine to form the single most basic, important, primal, and potentially traumatic occurrence of war," including genocide.[28] Studies have also found that soldiers across cultures may either *not* fire their weapons in combat, or may deliberately shoot above the enemy. Grossman concludes that "there can be no doubt that this resistance to killing one's fellow man is there and that it exists as a result of a powerful combination of instinctive, rational, environmental, hereditary, cultural, and social factors."[29]

So how, in warfare or genocide, do humans kill other humans? How does this most extreme form of direct violence happen? The answer, it appears, lies in a fundamental geographic concept, namely, that of distance. Indeed, it becomes readily apparent that a crucial component of the *geographical imagination* surrounding genocide and mass violence lies in the particular "act" of killing itself. Turning again to Grossman, we see that physical distance is extremely important for our understanding the behavior of killing. As the distance between perpetrator and victim increases, it becomes easier and less traumatic to kill. This, as we will see, accounts for the continued search for more efficient (and less traumatic) forms of mass murder in Nazi Germany.

Grossman notes that at "maximum range"—a range at which the killer is unable to perceive his or her individual victims without using some form of mechanical assistance (e.g., binoculars, radar, remote camera)—the act of killing is remarkably simple.[30] As the physical range decreases, however, killing becomes more difficult. Grossman finds that at "long range" (e.g., sniper weapons, tank fire) there begins to appear some disturbance at the act of killing. At midrange, a distance at which the soldier can see and engage the enemy with rifle fire but is unable to perceive the extent of the wounds inflicted or the sounds and facial expressions of the victim, there is an increased emotional toll. Killing at this range, according to Grossman, is often described as reflexive or automatic, and that the soldier experiences a range of emotions, from an initial feeling of euphoria or elation, followed by a period of guilt and remorse.[31]

Killing becomes most difficult at close range. Here lies "the undeniable certainty of *responsibility* on the part of the killer" (emphasis added).[32] Grossman concludes that at "close range the resistance to killing an opponent is tremendous. When one looks an opponent in the eye, and knows that he is young or old, scared or angry, it is not possible to deny that the individual about to be killed is much like oneself."[33] This accounts for the grim knowledge that during World War II German officers "experimented" with different techniques to minimize the trauma associated with close-range *mass murder*. Michael Burleigh explains that "one way of retaining a veneer of apparent decency was to kill in an apparently orderly military fashion." He continues that some officers "attempted to give murder a patina of legality by pronouncing sentences no court had delivered." Even these pronunciations were ineffective in reducing the moral repugnance of the task at hand: to systematically kill vast numbers of unarmed men, women, and children. Consequently, attempts were made to physically affect the spatial logic of killing. Burleigh elaborates:

> They started by lining up their victims in front of the graves in the manner of firing squads, but in too many cases this required the SS officers to deliver the *coup de grace* with their pistols. Next, they tried shooting standing victims from close behind, only to find that brains, blood and bits of skull flew back into their faces. After technical discussions in earshot of the next line of victims, it was deemed easier to shoot people kneeling or lying down within the graves, which minimized the first difficulty while making for a tidier body disposal. . . . Shooting people in a trench also prevented victims from leaping in and feigning death, a problem that had occurred with the firing-squad technique. Killing children raised the dilemma of whether to shoot the child or mother first, generally resolved in favor of killing the child first, since the traumatized mothers would be less trouble than a hysterical child and might even be relieved that their child did not have to see them die.

Burleigh concludes that "killing people became a job of work in which the killer could take a craftsman's pride. This shortened the moral distance, and enabled killing to become routine."[34]

This "geography of killing" has important implications for our broader understanding of killing as human behavior, particularly as it relates to the killing by "ordinary" people in the context of genocide and mass violence. Soldiers, we might argue, are trained to kill and thus are "better" equipped to overcome humanity's resistance to killing—although even here, we note,

Figure 1.2.   A member of an *Einsatzgruppe* prepares to shoot a Ukrainian Jew kneeling on the edge of a mass grave. *Source:* Sharon Paquette, courtesy of the United States Holocaust Memorial Museum

many German soldiers during the Holocaust were *unable* to carry out the bloodshed. But what about those who are not soldiers?

Whether one considers the Holocaust or the genocides in Cambodia, Rwanda, Darfur, and elsewhere, one cannot escape the observation that many killings were conducted by ordinary people—not trained soldiers.[35] Nor can we conclude that the perpetrators of mass violence were psychopaths or "natural born killers." As Christopher Browning writes, for example, "the Holocaust took place because at the most basic level individual human beings killed other human beings in large numbers over an extended period of time."[36] Such sustained killings throughout the Holocaust and other settings by ordinary people were instead the result of many factors: a broader context in which killings were *legally* permitted and *sanctioned* by state authorities, an organizational structure that facilitated killing, and the availability of weapons. For our present purposes, a crucial component is that of psychological distance—a morally informed geographical imagination.

Chirot and McCauley explain that "most humans have a sense of fairness that governs relations with others."[37] Consequently, physical distance—while important—must be tempered with an additional component. Distance is not simply physical; it also entails a social component. As Daly and Wilson write, "People who kill in spite of the inhibitions and penalties that confront them are people moved by strong passions."[38] These passions may be (and frequently are) intensely personal; but they also may be exceptionally social and political. A person's passion to kill may arise, paradoxically, from a broader "desire to *build a world* without conflict or enemies" (emphasis added).[39] In other words, the moral justification to kill another human being (or beings) may be predicated on the belief that such violence is necessary to obtain a particular result, such as a utopian society free from impurities. It was Mao Zedong who explained, in reference to the economic policies initiated in China during the Great Leap Forward (1958–1962), that "half of China may well have to die. If not half, one-third, or one-tenth—50 million—die."[40]

All human societies moralize and thus share basic categories such as obligatory, permitted, or forbidden actions.[41] Of significance is that, with a possible few exceptions, not all people within a given society are necessarily treated equally. Helen Fein refers to these social spaces as a "universe of obligation." She explains that there are some people who "must be taken into account, to whom obligations are due, by whom we can be held responsible for our actions."[42] Consequently, as Gamson elaborates, once people are defined as being outside that universe, offenses against them are not violations of the normative order and do not trigger criminal sanctions.[43] They may, in other words, be killed or allowed to die with impunity.

Moral inclusion *and* exclusion are therefore pivotal to our understanding of the geographies of violence in that these establish parameters around social relations. Susan Opotow explains that "inclusion in the scope of justice means applying considerations of fairness, allocating resources, and making sacrifices to foster another's well-being." Conversely, moral exclusion rationalizes and excuses harm inflicted on those outside the scope of justice. To exclude others from the scope of justice is to view them as being unworthy of fairness, resources, or sacrifice, and to see them as being expendable, undeserving, exploitable, or irrelevant.[44] In short, moral exclusion works to *legitimize* violence. Moral exclusion works against the reticence of taking another person's life. To morally exclude another human is to pave the way to kill that person. To morally exclude is to provide a calculated value of a life's worth.

The idea of moral exclusion is particularly important for our subsequent discussion of geographical imaginations and state-sanctioned violence. The geographic component of moral exclusion is identified as the *extent* of moral exclusion and refers to the scope of collective inclusion or exclusion. According to social psychologists, the *relational* practice of "us-them" thinking originates with social categorizations and these mental constructs—man/woman, black/white, citizen/noncitizen—are cognitive tools that segment, classify, and order our social environment.[45] The process of social categorization is thus foundational to the evocation of place. As James Waller explains, "Not only do social categorizations systematize our social world; they also create and define *our place in it*" (emphasis added).[46]

Waller suggests also that the use of social categorizations in assigning people to populations (and to places) has four salient effects: assumed similarity, out-group homogeneity, accentuation, and in-group bias.[47] These effects, moreover, are explicitly geographic. First, people who identify themselves as part of an in-group—defined, for example, by national origin—tend to perceive other in-group members as more similar than out-group members; these latter individuals, not surprisingly, are often classified as "foreign." Second, people perceive members of out-groups as all alike; generalizations, furthermore, are often based on one or two members. Third, perceived differences between in-groups and out-groups tend to be accentuated, or exaggerated. Finally, the mere act of dividing people into groups inevitably sets up a bias in group members in favor of the in-group and against the out-group.

Underlying these four effects is a process Kathleen Taylor defines as the "essence trap." According to Taylor, this "involves the *imagining* that everyone has a core character, the essence of who they are" (emphasis added).[48] Significantly, these essences are frequently portrayed as natural and invariable. The Tutsis in Rwanda, for example, were conceived as alien Others,

as were Jews in Nazi Germany. The process of social (and spatial) categorization, however, does not proceed based on natural divisions of humanity.
Social categories do not simply *include* groups; rather, the relational process
of categorization *produces* groups—this is a critical distinction, in that group
membership forms a cornerstone of the legal definition of genocide.

A common approach to justify the exclusion (and killing) of others, briefly
alluded to above, is to dehumanize the other. Simply stated, dehumanization
is a composite psychological practice that permits people to regard others as
being less than human. Waller explains that dehumanizing practices facilitate
the subsequent practices of exclusion, discrimination, and, ultimately, violence. He continues that once dehumanized, one's body "possesses no meaning. It is a waste, and its removal is a matter of sanitation. There is no moral or
empathetic context through which the perpetrator can relate to the victim."[49]
Hence, the practice of dehumanization serves to increase the psychological and relational distance between killer and victim. Such a dehumanization practice is readily seen in the rhetoric and propaganda of many (all?)
genocides and occurrences of mass violence. Waller observes, for example,
that "in the Holocaust . . . the Nazis redefined Jews as 'bacilli,' 'parasites,'
'vermin,' 'demons,' 'syphilis,' 'cancer,' 'excrement,' 'filth,' 'tuberculosis,'
and 'plague.' In the camps, male inmates were never to be called 'men' but
Haftlinge (prisoners), and when they ate the verb used to describe it was
fressen, the word for animals eating. Statisticians and public health authorities frequently would list corpses not as *corpses* but as *Figuren* (figures or
pieces), mere things. . . . Similarly, in a memo of June 5, 1942, labeled 'Secret
Reich Business,' victims in gas vans at Chelmno were variously referred to as
'the load,' 'number of pieces,' and the 'merchandise.'"[50]

Dehumanization constitutes a justification system within one's beliefs that
destroying an inherent evil is not the same as killing a human being. People
whose ordinary reality contains sharp inhibitions against inflicting violence
(or preventing the violence enacted by others) may switch to an alternative
reality—a different geographical imagination—that permits violence. Indeed,
it becomes clear that a geographical imagination permeates the practices of
genocide and mass violence. Semelin succinctly summarizes this point: humanity's ability to inflict pain and suffering on others is "mainly born out of
a mental process, a way of seeing some 'Other' being, of stigmatizing him,
debasing him, and obliterating him before actually killing him."[51] In other
words, our imagination empowers us "to see beyond the actual to the possible."[52] This includes the ability to envision a world *without* others, a world
"purified" of unwanted or undesirable others.

A geographic approach to genocide thus recasts the relevant questions to
be asked: Who, or which group, is granted or denied access to certain places?

What activities are deemed appropriate or not? What relations of power are maintained when "place" is invoked? And who has the authority, the ability, to define (and enforce) those places? In effect, the processes leading to moral inclusion or exclusion have a *geographic* component, one that is infused with power—a power we might phrase as *a will to space*. In particular, geographies become sites of contestation, the locus of social control where ideologies of racism, sexism, classism, and so forth are enacted.[53] Consequently, there may emerge a *moral* obligation (or perceived necessity) among some individuals to maintain certain places against the intrusion of unwanted Others. And when policing those places, it matters little whether the perceived intrusion or transgression was intentional or not.

Thus far, my discussion of violence has focused almost exclusively on individual "acts" of violence—the beatings, rapes, and killings that accompany genocide. However, there are other, more pervasive forms of violence, practices that not only facilitate individual acts of violence but are also considered violent in their own right. Here, I draw attention to *structural violence*.

Direct violence, as Susan Opotow explains, "is immediate, concrete, visible, and committed by and upon particular people."[54] This contrasts with structural violence, which "occurs as inequalities are structured into a society so that some have access to social resources that foster individual and community well-being—high quality education and health care, social status, wealth, comfortable and adequate housing, and efficient civic services—while others do not." Consequently, "structural violence normalizes unequal access to political and economic resources" in favor of one group over another group.[55] Through state policies and practices, unequal social relationships become cemented into political, economic, and social structures. This is seen, for example, in the differential access of some people to citizenship, employment, health care, and education. These embedded structures are considered violent in that a lack of access may quite literally mean the difference between life and death. Moreover, state policies and practices often provide the institutional legalization for violence that is experienced in genocides.

In short, while humans may be exceptionally violent, they are not necessarily prone to violence. Humans must be socialized to engage in (or at minimum, to tolerate) these acts—especially on a mass scale—and part of this socialization lies in the purpose of killing. Turning again to Gilligan, we understand that "all violence is an attempt to achieve justice."[56] Consequently, appeals to justice and legitimacy—such as the dehumanization of the victim—must necessarily increase as the physical and psychological distance of killing decreases. Were it not for the moral context established to *sanction* mass violence, it is questionable whether genocides could be carried out.

THE SOVEREIGN SPACES OF VIOLENCE

It is worth remembering that those "ordinary" people who committed the most atrocious acts during the Holocaust—or the genocides in Rwanda, or in Cambodia, or countless others—could have acted otherwise. Those ordinary people who stood by and said nothing, who did nothing, likewise could have acted otherwise. When confronted with the choice to do harm or to not do harm; to cause suffering or to not cause suffering; to ease suffering or to not ease suffering; humans have the capacity to make choices. However, these decisions are made within the confines of both political power and cultural norms.

Most acts of violence, including structural policies and practices, are instrumental; and most acts of violence are *justified* in the mind of the perpetrator or the bystander. On this notion, Hannah Arendt is clear: "Violence is by nature instrumental; like all means, it always stands in need of guidance and justification through the end it pursues."[57]

That violence, including mass violence, is instrumental—and rationalized—is an uncomfortable, but important, insight. As Svendsen writes, "Every desire may have something good about it, if only for the agent himself [*sic*], and even if the desire itself can be regarded as evil." Thus, for example, rape and murder are both considered evil, violent acts; yet *from the standpoint of the perpetrator, these acts have a good side.*[58] To be sure, as Primo Levi writes, "For anyone who reads these justifications the first reaction is revulsion."[59] Levi elaborates that such justification, expressed "in different formulations and with greater or lesser arrogance . . . in the end they substantially all say the same things: I did it because I was ordered to; others (my superiors) have committed acts worse than mine; in view of the upbringing I received, and the environment in which I lived, I could not have acted differently; had I not done it, another would have done it even more harshly in my place."[60]

All violence must be justified, either to oneself or to the group. In short, there must be a will to commit (or tolerate) violence. Consequently, to choose one course of action over another, including the choice to do violence, brings us to the thorny question of "*free will.*" Following Svendsen, most of our everyday activities are conducted without reflection; these behaviors are the habitual routines of quotidian life. However, to the extent that actions are caused by reflection—whereby one may identify a grounds for action—the grounds for action are related to the realization of a particular objective or desired goal.[61] In other words, the will to violence is (mostly) a reflective decision—it is based on a consideration of one's relationship toward another. Action, moreover, is not synonymous with wants, desires, feelings, or even needs. We may want something; we may feel envy or jealousy; we may also perceive that an injustice has occurred. But these "feelings" are not necessarily acted upon. Free will, according to Thomas Pink, is exercised through action—through what we deliberately do or refrain from doing.[62]

To be sure, the capacity to act is both partial and situated; we do not operate with complete "freedom" of choice. This is not to condone or to absolve those who participate, both directly and indirectly, in genocide, but to recognize the broader *political* context of one's actions. Most actions are bound up in a complex assemblage of sovereign powers that greatly inform both our decisions *to act* and our decisions *to not act*.

Sovereign power is commonly understood as "state power"; this is seen, for example, in Germany's decision to invade Poland in 1939. And throughout this and subsequent chapters, I will discuss in some detail the relation between sovereignty and genocide. First, however, it bears mentioning that sovereignty itself is scalar; that sovereignty "operates" at many levels of analyses. An exclusive focus on grand-scale examples of power, for example, may tend to overshadow our appreciation of the more mundane instances of state intervention into matters of life and death. Here, the work of Joe Painter is exceptionally informative. Recently, Painter has drawn attention to the "prosaic manifestations of state processes" and "the ways in which everyday life is permeated by stateness in various guises." In other words, Painter highlights the "mundane *practices* through which something we label 'the state' becomes present in everyday life."[63]

The pervasiveness of state intervention into the spatiality of life and death is well illustrated by Painter. He observes: "Giving birth, child rearing, schooling, working, housing, shopping, travelling, marrying, being ill, dying and countless other activities all involve us, to a greater or lesser extent, in relations with state institutions and practices, often in ways that are so taken for granted they are barely noticeable." As an example, Painter notes that in the United Kingdom, "by law a midwife must attend the birth of every child. The birth must be registered within six weeks and the mother and baby visited at home by a 'health visitor' (a state employee) to ensure that the child is well cared for. Until the child is sixteen, his or her primary carers will receive a fortnightly payment from the state to assist with the cost of its upbringing. By law every child must receive an education between the ages of five and sixteen, which for most children involves attending a state-funded school to learn a government-prescribed curriculum. As the child grows up, the state decides at what age and under what circumstances he or she can work for money, be held responsible for committing a crime, have sex, consume intoxicating substances, drive a motor vehicle, marry, vote and go to war."[64] To this end, Sandra Bartky refers to the disciplinary power that is both everywhere and nowhere; a situation whereby the "sovereign" is both every-one and no one.[65]

Sovereignty, as a legal system of authority, does have a *locus* however, and that location is in the modern state. As Robert Jackson explains, "sover-eignty is at the center of the political arrangements and legal practices of the modern world."[66] Consequently, although Bartky and others may note that the sovereign, in practice, may be nebulous (i.e., what are we actually saying when we ascribe some action to "the state"?), the fact remains that we all live within the confines of "the state." Again, turning to Jackson: "People may have some choice in determining [in] which state they will live—but they have no choice except to live in one state or another."[67] It is for this reason that questions of *state sovereignty* must factor into our study of genocide and mass violence.

To begin, we should acknowledge that *sovereignty* is a much used and much contested term. A standard definition holds that sovereignty refers to the "ultimate authority to rule within a polity."[68] Accordingly, sovereignty suggests two geographic components as it relates to territory. The first re-lates to internal sovereignty—the authority to rule within a delimited territo-rial state; a second refers to external sovereignty—the right of a sovereign government to rule its territory without external interference.[69] However, as Dahlman and others acknowledge, the actual practice of a territorial-based understanding of sovereignty is decidedly more slippery; sovereignty, in fact, is rarely so clear-cut or absolute. Stuart Elden, for example, notes that when a state finds itself failing to exercise one of the standard definitions of sovereignty, it can be said that its sovereignty has been rendered contingent.[70]

Although the details are subject to debate, it is widely agreed that the modern system of territorial sovereign states emerged during the fifteenth and sixteenth centuries. More precisely, sovereignty "is a historical innova-tion of certain Europe political and religious actors who were seeking escape from their subjection to the papal and imperial authorities of medieval Eu-rope and to establish their independence of all other authorities, including each other."[71] As John Agnew explains, the older hierarchical arrangements in Europe, consisting of the Roman Church and the Holy Roman Empire, feudal obligations, and theological claims to just rule, gradually dissolved, to be replaced by a system of territorial states.[72] Universal ideals predicated on religion and empirical hierarchies gave way to separate states, each capable of defining its own goals and objectives. With respect to the Church and Empire, for example, Michel Foucault elaborates that these "two great forms of universality . . . had no doubt become a sort of empty envelope, an empty shell." He continues that although these institutions still retained their power of focalization, attraction, and intelligibility, they also had lost their vocation and meaning, at least at the level of this universality.[73]

The gradual transition to an international system of sovereign states entailed a radical reconfiguration of political geography. Medieval Europe consisted of various *regna*, or "islands of local political authority scatted across the western part of the former Roman Empire." Such *regna* included the Visigoths, Huns, Ostrogoths, Vandals, Franks, Saxons, and so on. In time, particular *regna* came to be identified with specific peoples and territories; it is within the *regna* that the idea of "nationality" emerged.[74]

According to Jackson, many *regna* consisted of heterogeneous populations, oftentimes dispersed across noncontiguous territories. Social and political relations were personal, not territorial, and reflected a complex arrangement between kings and cardinals, barons and bishops. In short, the "political world of the Middle Ages was diverse, dislocated, and disjointed," but one that also reflected deeply the "interdependency and involvement of both church and state."[75]

This complex political system changed as a result of the Protestant Reformation and the subsequent Catholic Counter-Reformation. Internecine struggles between kings and popes—a separation of church and state (but not God and state)—gave rise to a new geographical imagination, one in which rulers and their subjects expressed loyalty *not* to the church, but rather to the "state." New ideas, manifest in the writings of Machiavelli, Martin Luther, Jean Bodin, and Thomas Hobbes, provided legitimacy to the evolving state concept. In time, the "state" would monopolize "all legal authority by incorporating a sovereign with a lawful right to command his or her subjects without interference from emperors or barons, popes or other prelates, either outside the country or inside."[76]

From the sixteenth century onwards, therefore, relations between states no longer were perceived in the form of princely rivalries, but rather in the form of national competition. This was competition within a territorial state system. Indeed, one might surmise that "loyalty" or "obligation" or "deference" to the Holy Roman Empire or the Church was transferred to the concept of the territorial state system. As Foucault suggests, if states were to exist alongside each other in a competitive relationship, a system had to be found that would limit the ambitions and growth of other states as much as possible, while still leaving each state sufficient opportunities to maximize its own growth without provoking its adversaries and without leading to its own disappearance.[77] In other words, it was imperative that states work to ensure the continuation of competition while simultaneously precluding the imposition of any factor that might disrupt or dissolve the overall state system. In fact, we continue to shoulder this legacy, as witnessed by the reluctance of sovereign states to interfere in the affairs of other sovereign states—even if such intervention might "prevent" a genocide.

Concerns over the territorial integrity of states *and* the integrity of the territorial state system became (and continue to be) paramount. As Hubert Dreyfus and Paul Rabinow explain, the history, geography, climate, and demography of a particular *state* became more than mere curiosities. These became "crucial elements in a new complex of power and knowledge and, subsequently, the development of academic fields such as demography, sociology, and geography. The government, particularly the administrative apparatus, needed knowledge that was concrete, specific, and measurable in order to operate effectively. This enabled it to ascertain precisely the state of its forces, where they were weak and how they could be shored up."[78]

For our present purposes, the most salient aspect of sovereignty lies in the relationship between the "territorial state" and its "population." In other words, our focus is necessarily on the practices—the calculations—by which governments rule and regulate, discipline and control, the populations within their territorial domains. As Crampton and Elden write, "Forms of organizing, conceptualizing and managing the population can be seen in technologies such as the census and representational discourses, statistics, planning and cartography, as well as political expressions such as geopolitics, government and colonial ordering."[79] However, these calculations may also be seen, as detailed later, in the sovereign state's "monopoly of legitimate physical violence"—the state's right to intervene in matters of life and death.[80]

The concept of "population" is explicitly tied to the development of the modern state, in that population emerged as a factor of state strength, to be considered and calculated alongside other state characteristics (e.g., finances, resources, and territorial size). The timing was not happenstance. The elaboration of a concept of population was a gradual process that was both technical and theoretical, relying on the development of statistics and census taking, and the techniques of epidemiology, demography, and political philosophy.[81] This has significant implications for the historical and comparative study of genocides, in that the political *meaning* of population as a concept was fundamentally transformed. As Thomas Lemke writes, "population constitutes the combination and aggregation of individualized patterns of existence to a new political form."[82] With the emergence of the territorial state, populations were no longer conceived as the simple sum of individuals inhabiting a territory; instead, populations were conceived as a technical-political object of management and government. The relationship between the sovereign and the population is therefore not simply one of obedience or the refusal of obedience. Rather, "populations" become productive through state interventions—through a series of techniques, practices, and calculations.[83]

Conceptually, the idea of populations as a collective of bodies constituting some kind of definable unit to which measurements pertain emerged in the sixteenth century.[84] This is seen, for example, in the writings of John Graunt, William Petty, William Farr, Johann Peter Süssmilch, and Thomas Malthus, among others. Graunt for example identified certain regularities within populations: that child mortality is always higher than adult mortality; that populations exhibit a slighter higher proportion of male births compared to female births, but that male mortality is higher, thus leading to a more equal proportion of boys and girls. Süssmilch, likewise, demonstrated the existence of certain statistical regularities in population data. While searching for a divine order, or evidence of God's planning, Süssmilch—who was a clergyman and published a book titled *The Divine Ordinance Manifested in the Human Race through Birth, Death, and Propagation*—examined masses of demographic data. In his quest for regularities, Süssmilch discerned the balance of births and deaths and subsequently produced a life table; this knowledge was in fact used for actuarial purposes well into the nineteenth century.[85] That Süssmilch was able to conduct these analyses was itself a function of a growing state presence, for it was during the Renaissance that governments explicitly and consistently undertook the collection of population data, and that statistical studies based on these data were rapidly improved.[86]

The modern concept of population became dependent on the establishment of practical equivalences among political subjects—the people within a state.[87] The concept of "population" in other words hinged on aggregates and regularities. Adrian Bailey notes that "essentialist approaches to populations as groups not only infer group meaning from component characteristics, but regard these components as fixed, naturalized and stable elements."[88] Curtis elaborates:

> As an object of knowledge, population is primarily a statistical artifact. The establishment of practical equivalences means that population is connected to the law of large numbers, which causes individual variation to disappear in favor of regularity. In its developed forms, population is bound up with the calculus of probabilities. Population makes it possible to identify regularities . . . and such things may be both analytic tools and objects of intervention, such as birth, death, or marriage rates.[89]

What is most remarkable is how the concept of population would inform *political practice*. Again turning to Curtis, we see that beginning in the eighteenth century, the consolidation of population as a *component* of the state "depends on the establishment of equivalences among the subjects within a particular territory"; consequently, political-scientific knowledge "depends on the discipline of potential objects of knowledge. It is only on the grounds of constructed and enforced equivalences that one body comes to equal another, that each death, birth, marriage, divorce and so on, comes to be the equivalent of any other." Curtis concludes that "it is only on the grounds of such constructed equivalences that it is possible for statistical objects to emerge in the form of regularities and to become the objects of political practice. Population is coincident with the effective capacity of sovereign authority to discipline social relations."[90] From this point forward it became possible to speak of a state's *population*, as if that population had a transcendental existence and experience above and beyond the government. Furthermore, it is here that we can locate the construction of populations into subgroups that contribute to or retard the general welfare and life of the population as a whole.[91] For Dean, it is this proclivity that led to the discovery among the population *writ large* of criminal classes, for example, or the feebleminded and the imbeciles, the inverts and the degenerates, the unemployable and the abnormal.[92]

The study of populations coincided with an ideological shift concerning the nature of governance—a shift that made possible the elaboration of distinctively liberal governmental techniques and rationalities.[93] Governments were no longer primarily concerned with the proper distribution of things, by which Dean means an ordering and regulation of humans in their relations to various heterogeneous entities (for example, wealth, industry, land) and orderly settlement and movements between and within territories. Rather, "the point at which population ceases to be the sum of the inhabitants within a territory and becomes a reality *sui generis* with its own forces and tendencies is the point at which this dispositional government of the state begins to meet a government through social, economic, and biological processes."[94]

The concept of population introduces several key elements that have broad ramifications for the art of governance—and, consequently, for the study of genocide and mass violence. First is that the idea of population introduces a different conception of the governed. The members of a population are no longer subjects bound together in a territory who are obliged to submit to their sovereign. Instead, they are also conceived as living and working social beings, with their own customs, habits, and histories. Second, populations are defined in relation to matters of life and death, health and illness, propagation and longevity, all of which can be *known* by statistical, demographical, and epidemiological instruments. Third, the concept of population impacts the idea of a collective entity, the knowledge of which is irreducible to the knowledge that any of its members may have of themselves. In other words, the "population is not just a collection of living, working and speaking subjects; it is also a particular objective reality of which one can have knowledge."[95]

Supervision of population processes—such as marriage, births, deaths—was effected through a series of *state* interventions Foucault terms "biopolitics." *Biopolitics* entails "a set of processes such as the ratio of births to deaths, the rate of reproduction, the fertility of a population, and so on. It is these processes—the birth rate, the mortality rate, longevity, and so on—together with a whole series of related economic and political problems . . . [that] become biopolitics' first objects of knowledge and the targets it seeks to control."[96] The individual, to the extent that bodies were considered, was "of interest exactly insofar as he [*sic*] could contribute to the strength of the state. The lives, deaths, activities, work, miseries, and joys of individuals were important to the extent that these everyday concerns became politically useful."[97]

Biopolitics can be seen as an extension of the state onto individual bodies. This is not to say, however, that the body previously had been separate from political oversight. Many of the writings of Michel Foucault demonstrate quite graphically the various disciplinary techniques by which bodies were subjected to political surveillance and control. The subjected body, according to Foucault, "is directly involved in a political field: power relations have an immediate hold upon it; they invest it, mark it, train it, torture it, force it to carry out tasks, to perform ceremonies, to emit signs."[98] He continues:

> This political investment of the body is bound up in accordance with complex reciprocal relations, with its economic use: it is largely as a force of production that the body is invested with relations of power and domination; but, on the other hand, its constitution as labour power is possible only if it is caught up in a system of subjection (in which need is also a political instrument meticulously prepared, calculated and used); the body becomes a useful force only if it is both a productive body and a subjected body.[99]

Discipline is a type of power, one that Foucault terms an "anatomo-politics" of the human body. As explained by Foucault, "it is always the body that is at issue—the body and its forces, their utility and their docility, their distribution and their submission."[100] Moreover, it is a normalizing practice, consisting of "a whole set of instruments, techniques, procedures, and levels of applications."[101] Disciplinary techniques of normalization impose homogeneity. The individual becomes the object and effect of various techniques and practices, and becomes the measure of the homogenized norm against which disciplinary tactics can be brought to bear.[102] In both Nazi Germany and Cambodia, for example, specific nonconforming bodies were subject to intense state surveillance and violence. However, the homogenizing processes establish also spaces from which any deviation from the norm may be identified and corrected—if not eliminated. Thus, a person perceived to be incapable of work within Nazi Germany might be sent to the gas chambers, while in Cambodia those not conforming to party ideologies could be tortured and executed as a "traitor." Moreover, not all bodies, from the standpoint of the state, are able (or deemed worthy enough) to be "normalized." As James writes, "Under Nazi Germany, those designated biologically abnormal—the handicapped body, the gay body, the Jewish body, the Gypsy body, the communist or socialist body, and the black body—were to be sterilized, euthanized, or eradicated; considered defective or a subspecies of humans, they could not be normalized under state ideology."[103]

Unlike the disciplining of individual bodies, biopolitics addresses the population, with populations (not bodies) as the political-economic target. New technologies of power—including population forecasts, statistical estimates, and various other demographic measures and concepts—were applied not to individual bodies, but to collectivities and aggregates or, simply, populations. And the purpose of these techniques, according to Foucault, was not to modify any given phenomenon as such, or to modify a given body insofar as he or she was an individual, but to intervene at the level at which these general phenomena are determined.[104] For example, the "mortality rate" or the "birth rate" was to be modified. This could be accomplished through various regulatory mechanisms that operated at larger, aggregate scales.

The distinction between these two techniques of power—anatomo-politics and biopolitics—has a tremendous bearing on our understanding of genocide. One technique, the anatomo-politics, is disciplinary and centers on the body. This exercise of power produces individualizing effects and manipulates the body as a source of forces that have to be rendered both useful and docile. The second set brings together the mass effects characteristic of a population, and attempts to control the series of random events that can occur in a living mass (for example, births, deaths, and illnesses).[105] Biopolitics therefore does not exclude considerations of the anatomo-politics of the human body. Rather, the two forms of power are seen as complementary; they exist on different levels, one directed toward the body, the other toward the population. The combination of the two contributes to a particular era of "biopower," marked by an explosion of numerous and diverse techniques for achieving the subjugation of bodies and the control of populations.

LIFE, DEATH, AND SOVEREIGNTY

The principal expression of state sovereignty resides, to a large degree, in the power and the capacity to dictate who may live and who must die.[106] According to Adam Thurschwell there is a remarkable consensus in Western political philosophy, from the beginnings of post-Westphalian modernity to the present day, that identifies the essence of political sovereignty with the sovereign power to kill—or, to be more precise, the sovereign's right to claim the death of the citizen, whether in the form of direct killing or in the form of demanding that the citizen sacrifice his or her own life for the life of the state by submitting to military conscription.[107] In *Leviathan*, for example, Thomas Hobbes defines the sovereign as "he [*sic*] who alone retains the natural "right" to everything, and to do whatsoever he thought necessary to his own preservation; subduing, hurting or killing any man in order thereunto" while Jean-Jacques Rousseau, in *The Social Contract*, asserted that the citizen's "life is not now, as it once was, merely nature's gift to him [*sic*]. It is something that he holds, on terms, from the State."[108]

However much these philosophical traditions might differ in the normative conclusions that they draw about the exercise of political power and the concomitant restraints on the exercise of power, they all posit "political power" itself—sovereignty as such, prior to any normative evaluation—in essentially the same way: as the sovereign's right, on the one hand, to kill its citizens (and others on its territory as well) and, on the other, to conscript its citizens to fight and die in its own defense."[109]

Following Michel Foucault, in the classic conception of sovereignty, the right of life and death was one of the sovereign's basic attributes. In other words, to say that "the sovereign has a right of life and death means that he [*sic*] can . . . either have people put to death or let them live." Life and death, therefore, are removed from the realm of the "natural" and fall within the field of governance. Foucault suggests, also, that the sovereign cannot grant life in the same way that he or she can inflict death. The right of life and death, therefore, "is always exercised in an unbalanced way: the balance is always tipped in favor of death." Consequently, the "very essence of the right of life and death is actually the right to kill: it is at the moment when the sovereign can kill that he [*sic*] exercises his right over life." [110]

The origination and transformation of the modern territorial state coincided with a changed ethic regarding the sovereign's right over life and death. According to Foucault, the ancient right to take life or to let live was gradually replaced by a power to foster life or to disallow life to the point of death.[111] This shift, which paralleled the rise of modern medicine and its attendant body of medical ethics, entailed a repositioning of the population vis-à-vis the state. As Thurschwell questions, how could the sovereign power exercise its highest prerogatives by putting people to death, when its main role was to ensure, sustain, and multiply life, to put this life in order?[112] The answer lies in the emergence of what Foucault termed *biopower*—a development that is now well entrenched within geography and the social sciences more broadly.

As we have seen, beginning broadly in the seventeenth century, the sovereign power over life developed along two competing but complementary poles. On the one hand, there emerged an anatomo-politics of the human body, seeking to maximize the forces of the body and to integrate it into efficient, productive systems. On the other hand, there emerged a suite of regulatory controls, a biopolitics of the population. This latter development was imbued with the mechanisms of life in its totality: birth, morbidity, mortality.[113] As Foucault writes, the old power of death that symbolized sovereign power was now carefully supplanted by a "calculated management of life."[114] This is a key point—one that echoes in the words of the genocide scholar Irving Horowitz. While noting the "presumption of taking life as a unique capacity, morally and legally of the authority system within a society," Horowitz cautions that this is but half of the story. Thus, the other component—the "giving life"—is an equally significant function of state power.[115]

Of significance, also, is the observation that the modern state's right to "foster life or to disallow it to the point of death" never completely erased the classical right to kill. This is seen most immediately in the form of capital punishment; it appears also in matters of torture, political assassinations, and war. As Sarat and Culbert explain, "In a regime dedicated to putting and keeping life in order and safe, the state may still exercise the right to death associated with the classic sovereign. To do so, however, it has to describe those who will be put to death as incorrigible monsters or as biological hazards so that their demise and final disposal can be represented as an unpleasant but necessary task that the state reluctantly but decisively undertakes for the well-being of its citizens."[116] Again, this idea will be brought to the fore in our subsequent case studies of genocide and mass violence.

In short, what the modern state reveals is a decidedly more nuanced management of life and death than that proposed by Foucault. At this point it is appropriate to consider the extension of Foucault's work as developed by Giorgio Agamben. In a corrective to Foucault, Agamben argues that *thanatopolitics*—a politics of death—is actually the first principle of biopolitics. As Rose explains, a thanatopolitics is predicated on the understanding that "life itself is subject to a judgment of worth, a judgment that can be made by oneself (suicide) but also by others (doctors, relatives) but is ultimately guaranteed by a sovereign authority (the state)."[117] Agamben makes this argument through a focus on *homo sacer*—an obscure figure of archaic Roman law. For Agamben, *homo sacer* is one who can be killed with impunity, one whose death constitutes neither homicide nor sacrifice. *Homini sacri*, consequently, are situated outside both human and divine law; they are "included" in politics only through their "exclusion"; they constitute, in short, "bare life."[118]

This has immense importance, in that it directs attention to the bioethical propositions that are drawn upon in the practice of statecraft. Again, following Rose, "At the very moment when political sovereignty was established over a territory, power was linked to the living bodies of its subjects, if only because this is what enables the sovereign alone to make legitimate political use of their death."[119] As such, decisions to kill, to foster life, or to disallow life to the point of death are made within a context of state valuation: matters of life and death are understood within the domain of bioethics. Agamben writes, for example, that "in modern biopolitics, sovereign is he [*sic*] who decides on the value or the nonvalue of life as such."[120]

THE PATH AHEAD

"We can dream of peace," Paul Kahn writes, and "we can imagine a global order of perfect lawfulness." However, Kahn reminds us that "we dream of these things from a position deep within *the political formation of a state* that has its origins in violence, that will maintain itself through violence, and that claims a unique right to demand sacrifice of all of its citizens."[121] Those events designated as genocide, I suggest, are predicated upon a violent enforcement of the "will to space." Stated differently, my thesis is that *mass violence results from the imposition of state-sanctioned normative geographical imaginations that justify and legitimate unequal access to life and death.* And while much violence committed within genocidal events is direct (i.e., the rape and murder of individual bodies), I suggest further that both direct violence and the broader structures of violence are dependent upon state policies and practices. Thus, we must first and foremost direct our attention to that of the state. Following Horowitz, "Genocide is neither capricious nor accidental. Neither does it follow any inexorable 'laws' of economic development." Rather, it is "the special relationship between the state and its monopoly on life-taking propensities" that underwrites genocide and mass violence.[122]

As subsequent chapters demonstrate, modern states—including what we might term "genocidal states"—exhibit complex (and contradictory) practices that impinge on the health and well-being of individual bodies and populations. A geographical perspective on genocide highlights that mass violence, in the imaginations of perpetrators, is viewed as an effective—and *legitimate*—strategy of state-building. What the following episodes of mass violence—Germany, China, and Cambodia—reveal is how specific "states" articulated and acted upon particular geographical imaginations that determined the moral worth of lives that could be killed or disallowed life to the point of death. As Foucault writes, "And sometimes what he [*sic*] has to do for the state is to live, to work, to produce, to consume; and sometimes what he has to do is to die."[123]

NOTES

1. Robert N. Proctor, *Racial Hygiene: Medicine under the Nazis* (Cambridge, MA: Harvard University Press, 1988), 184.

2. David Dranove, *What's Your Life Worth? Health Care Rationing . . . Who Lives? Who Dies? And Who Decides?* (New York: Prentice Hall, 2003), 139.

3. Dranove, *What's Your Life Worth?*, 142.

4. David Hamburg, *Preventing Genocide: Practical Steps toward Early Detection and Effective Action* (Boulder, CO: Paradigm Publishers, 2008).

5. Hamburg, *Preventing Genocide*, 6.

6. Ben Kiernan, *Blood and Soil: A World History of Genocide and Extermination from Sparta to Darfur* (New Haven, CT: Yale University Press, 2007), 6.

7. Zygmunt Bauman, *Modernity and the Holocaust* (Ithaca, NY: Cornell University Press, 2000); Martin Shaw, *War & Genocide: Organized Killing in Modern Society* (Malden, MA: Polity Press, 2003).

8. Lennart Vriens, "Peace Education: Cooperative Building of a Humane Future," *Pastoral Care in Education* 15(1997): 25–30; at 27.

9. Angka (Angkar) was the secretive "Party Center" of the Communist Party of Kampuchea (CPK). This organization, as well as the Cambodian genocide, is developed more fully in chapter 3.

10. Roeun Sam, "Living in Darkness," in *Children of Cambodia's Killing Fields: Memoirs by Survivors*, compiled by Dith Pran and edited by Kim DePaul (New Haven, CT: Yale University Press, 1997), 73–81; at 76.

11. Sam, "Living in Darkness," 76.

12. Henri Lefebvre, *The Production of Space*, translated by D. Nicholson-Smith (Oxford, UK: Blackwell, 1991), 85.

13. Tim Cresswell, *In Place/Out of Place: Geography, Ideology and Transgression* (Minneapolis: University of Minnesota Press, 1996), 11.

14. Edward W. Soja, *Postmodern Geographies: The Reassertion of Space in Critical Social Theory* (New York: Verso, 1989), 80.

15. John K. Wright, "Terrae Incognitae: The Place of the Imagination in Geography," *Annals of the Association of American Geographers* 37(1947): 1–15.

16. Wright, "Terrae Incognitae," 1.

17. Edward Said, *Orientalism* (New York: Vintage Books, 1979); see also Bill Ashcroft and Pal Ahluwalia, *Edward Said* (New York: Routledge, 1999), 61.

18. Martin Lewis and Kären Wigen, *The Myth of Continents: A Critique of Metageography* (Berkeley: University of California Press, 1997).

19. Derek Gregory, "The Lightning of Possible Storms," *Antipode* 36(2004): 798–808; at 798.

20. Archeologists and anthropologists, among others, remain somewhat divided on this topic. Studies over the last three decades have revealed that numerous species routinely kill others of their own kind. Male hippopotamuses, for example, will kill the offspring of another male so that they may then mate with the female. And closer to home, so to speak, considerable research has documented the organized violence of chimpanzees. However, debate remains as to whether humans' propensity for large-

scale warfare is comparable to the small-scale raiding exhibited in chimpanzee politics. See, for example, Frans de Waal, *Chimpanzee Politics: Power and Sex among Apes* (Baltimore: Johns Hopkins Press, 1992); Matt Ridley, *The Red Queen: Sex and the Evolution of Human Nature* (New York: Harper Perennial, 2003); Frans de Waal, *Our Inner Ape: A Leading Primatologist Explains Why We Are Who We Are* (New York: Riverhead Books, 2005); Nicholas Wade, *Before the Dawn: Recovering the Lost History of Our Ancestors* (New York: Penguin Books, 2007).

21. Milton Leitenberg, "Deaths in Wars and Conflicts in the 20th Century," *Cornell University Peace Studies Program Occasional Paper #29*. See also Rudolph J. Rummel, *Death by Government: Genocide and Mass Murder in the Twentieth Century* (New Brunswick, NJ: Transaction Publishers, 1994); and Mark Levene, "Why Is the Twentieth Century the Century of Genocide?" *Journal of World History* 11(2000): 305–36.

22. Lars Svendsen, *A Philosophy of Evil* (Champaign, IL: Dalkey Archive Press, 2010), 30.

23. James Gilligan, *Violence: Reflections on a National Epidemic* (New York: Vintage Books, 1997), 9.

24. Daniel Chirot and Clark McCauley, *Why Not Kill Them All? The Logic and Prevention of Mass Political Murder* (Princeton, NJ: Princeton University Press, 2006), 51.

25. See, for example, S. L. A. Marshall, *Men Against Fire* (Gloucester: Peter Smith, 1978); Christopher Browning, *Ordinary Men: Reserve Police Battalion 101 and the Final Solution in Poland* (New York: Harper Perennial, 1992); and Dave Grossman, *On Killing: The Psychological Cost of Learning to Kill in War and Society* (New York: Back Bay Books, 1996).

26. Grossman, *On Killing*, 28.

27. See James A. Tyner, "Toward a Nonkilling Geography: Deconstructing the Spatial Logic of Killing," in *Toward a Nonkilling Paradigm*, edited by Joám Evans Pim (Honolulu, HI: Center for Global Nonkilling, 2009), 169–85.

28. Grossman, *On Killing*, 31.

29. Grossman, *On Killing*, 39.

30. Grossman, *On Killing*, 107.

31. Grossman, *On Killing*, 109–11.

32. Grossman, *On Killing*, 114.

33. Grossman, *On Killing*, 118.

34. Michael Burleigh, *Moral Combat: Good and Evil in World War II* (New York: Harper, 2011), 407–08.

35. See, for example, Browning, *Ordinary Men*; Alexander L. Hinton, *Why Did They Kill? Cambodia in the Shadow of Genocide* (Berkeley: University of California Press, 2005); Jacques Semelin, *Purify and Destroy: The Political Uses of Massacre and Genocide*, translated by Cynthia Schoch (New York: Columbia University Press, 2007).

36. Browning, *Ordinary Men*, xvii.

37. Chirot and McCauley, *Why Not Kill Them All?*, 53.

38. Martin Daly and Margo Wilson, *Homicide* (London: Transaction Publishers, 1988), 12.

39. Semelin, *Purify and Destroy*, 33.

40. Quoted in Jung Chang and Jon Halliday, *Mao: The Unknown Story* (New York: Anchor Books, 2006), 431.

41. Kathleen Taylor, *Cruelty: Human Evil and the Human Brain* (Oxford: Oxford University Press, 2009), 37.

42. Helen Fein, *Imperial Crime and Punishment: The Massacre at Jallianwala Bagh and British Judgment, 1919–1920* (Honolulu: University of Hawaii Press, 1977), 7.

43. William A. Gamson, "Hiroshima, the Holocaust, and the Politics of Exclusion," *American Sociological Review* 60(1995): 1–20; at 3.

44. Susan Opotow, "Reconciliation in Time of Impunity: Challenges for Social Justice," *Social Justice Research* 14(2001): 149–170; at 156.

45. James Waller, *Becoming Evil: How Ordinary People Commit Genocide and Mass Killing* (New York: Oxford University Press, 2002), 239.

46. Waller, *Becoming Evil*, 239.

47. Waller, *Becoming Evil*, 239–40.

48. Taylor, *Cruelty*, 9.

49. Waller, *Becoming Evil*, 245.

50. Waller, *Becoming Evil*, 246.

51. Semelin, *Purify and Destroy*, 9.

52. David Livingstone Smith, *The Most Dangerous Animal: Human Nature and the Origins of War* (New York: St. Martin's Griffin, 2007), 101.

53. James A. Tyner, *Space, Place, and Violence: Violence and the Embodied Geographies of Race, Sex, and Gender* (New York: Routledge, 2012).

54. Opotow, "Reconciliation," 151.

55. Opotow, "Reconciliation," 151.

56. Gilligan, *Violence*, 11.

57. Hannah Arendt, *On Violence* (New York: Harcourt, 1970), 51.

58. Svendsen, *Philosophy of Evil*, 85.

59. Primo Levi, *The Drowned and the Saved* (New York: Vintage Books, 1989), 26.

60. Levi, *Drowned and the Saved*, 26.

61. Svendsen, *Philosophy of Evil*, 109.

62. Thomas Pink, *Free Will: A Very Short Introduction* (Oxford: Oxford University Press, 2004), 6.

63. Joe Painter, "Prosaic Geographies of Stateness," *Political Geography* 25(2006): 752–74; at 753.

64. Painter, "Prosaic Geographies," 753.

65. Sandra Lee Bartky, *Femininity and Domination: Studies in the Phenomenology of Oppression* (New York: Routledge, 1990), 74.

66. Robert Jackson, *Sovereignty: Evolution of an Idea* (Malden, MA: Polity Press, 2007), ix.

67. Jackson, *Sovereignty*, x.

68. Carl T. Dahlman, "Sovereignty," in *Key Concepts in Political Geography*, edited by C. Gallaher, C. T. Dahlman, M. Gilmartin, A. Mountz, and P. Shirlow (Los Angeles: Sage Publications, 2009), 28–40; at 28.

69. Dahlman, "Sovereignty," 28; see also S. D. Krasner, "Problematic Sovereignty," in *Problematic Sovereignty: Contested Rules and Political Possibilities*, edited by S. D. Krasner (New York: Columbia University Press, 2001), 1–23.

70. Stuart Elden, "Contingent Sovereignty, Territorial Integrity and the Sanctity of Borders," *SAIS Review* 26(2006): 11–25; at 15.

71. Jackson, *Sovereignty*, 6.

72. John Agnew, "The Territorial Trap: The Geographical Assumptions of International Relations Theory," *Review of International Political Economy* 1(1994): 53–80; at 60.

73. Michel Foucault, *Security, Territory, Population: Lectures at the Collège de France, 1977–1978*, translated by Graham Burchell (New York: Picador, 2007), 291.

74. Jackson, *Sovereignty*, 25–26.

75. Jackson, *Sovereignty*, 32–33.

76. Jackson, *Sovereignty*, 38–39.

77. Foucault, *Security, Territory, Population*, 296–97.

78. Hubert L. Dreyfus and Paul Rabinow, *Michel Foucault: Beyond Structuralism and Hermeneutics* 2nd ed. (Chicago: University of Chicago Press, 1983), 107.

79. Jeremy Crampton and Stuart Elden, "Space, Politics, Calculation: An introduction," *Social and Cultural Geography* 7(2006): 681–85; at 682.

80. See Max Weber, *Economy and Society: An Outline of Interpretive Sociology* (New York: Bedminster Press, 1968).

81. Mitchell M. Dean, *Governmentality: Power and Rule in Modern Society* (Thousand Oaks, CA: Sage, 1999), 108.

82. Thomas Lemke, *Biopolitics: An Advanced Introduction*, translated by Eric F. Trump (New York: New York University Press, 2011), 37. See also Stephen Legg, "Foucault's Population Geographies: Classifications, Biopolitics and Governmental Spaces," *Population, Space and Place* 11(2005): 137–56; and Chris Philo, "Sex, Life, Death, Geography: Fragmentary Remarks Inspired by Foucault's Population Geographies," *Population, Space and Place* 11(2005): 325–33.

83. Foucault, *Security, Territory, Population*, 70–71.

84. John Caldwell, "Demographers and the Study of Mortality: Scope, Perspectives, and Theory," *Annals of the New York Academy of Sciences* 954(2001): 19–34; at 20.

85. Caldwell, "Demographers," 22; see also Preston E. James and Geoffrey J. Martin, *All Possible Worlds: A History of Geographical Ideas*, 2nd ed. (New York: Wiley, 1981), 102.

86. James and Martin, *All Possible Worlds*, 102–3.

87. Bruce Curtis, "Foucault on Governmentality and Population: The Impossible Discovery," *Canadian Journal of Sociology* 27(2002): 505–33; at 508.

88. Adrian Bailey, *Making Population Geography* (London: Hodder Arnold, 2005), 118.

89. Curtis, "Foucault on Governmentality," 508–9.

90. Curtis, "Foucault on Governmentality," 529.

91. Dean, *Governmentality*, 29.

92. Dean, *Governmentality*, 100.

93. Dean, *Governmentality*, 107.

94. Dean, *Governmentality*, 95–96.

95. Dean, *Governmentality*, 107.

96. Michel Foucault, *"Society Must Be Defended": Lectures at the Collège de France, 1975–1976*. Translated by David Macey (New York: Picador, 2003), 243.

97. Dreyfus and Rabinow, *Michel Foucault*, 139.

98. Michel Foucault, *Discipline and Punish: The Birth of the Prison*. Translated by Alan Sheridan (New York: Vintage Books, 1979), 25.

99. Foucault, *Discipline and Punish,* 25–26.

100. Foucault, *Discipline and Punish*, 25.

101. Foucault, *Discipline and Punish,* 215

102. Philip Barker, *Michel Foucault: An Introduction* (Edinburgh: Edinburgh University Press, 1998), 58.

103. Joy James, *Resisting State Violence: Radicalism, Gender and Race in U.S. Culture* (Minneapolis: University of Minnesota Press, 1996), 27–28.

104. Foucault, *"Society Must Be Defended,"* 246.

105. Foucault, *"Society Must Be Defended,"* 249.

106. Achilles Mbembe, "Necropolitics," *Public Culture* 15(2003): 11–40; at 11.

107. Adam Thurschwell, "Ethical Exception: Capital Punishment in the Figure of Sovereignty," in *States of Violence: War, Capital Punishment, and Letting Die*, edited by Austin Sarat and Jennifer L. Culbert (Cambridge: Cambridge University Press, 2009), 270–96; at 282.

108. Quoted in Thurschwell, "Ethical Exception," 283.

109. Thurschwell, "Ethical Exception," 283–84.

110. Foucault, *"Society Must Be Defended,"* 240.

111. Michel Foucault, *The History of Sexuality: An Introduction* (New York: Vintage Books, 1990), 138.

112. Thurschwell, "Ethical Exception," 274.

113. Paul Rabinow and Nikolas Rose, "Biopower Today," *BioSocieties* 1(2006): 195–217; at 196.

114. Foucault, *History of Sexuality*, 140.

115. Irving Louis Horowitz, *Taking Lives: Genocide and State Power*, 3rd ed. (New Brunswick, NJ: Transaction Books, 1980), xiii.

116. Austin Sarat and Jennifer L. Culbert, "Introduction: Interpreting the Violent State," in *States of Violence: War, Capital Punishment, and Letting Die*, edited by Austin Sarat and Jennifer L. Culbert (Cambridge: Cambridge University Press, 2009), 1–22; at 6.

117. Nikolas Rose, *The Politics of Life Itself: Biomedicine, Power, and Subjectivity in the Twenty-First Century* (Princeton, NJ: Princeton University Press, 2007), 57.

118. Giorgio Agamben, *Homo Sacer: Sovereign Power and Bare Life* (Stanford, CA: Stanford University Press, 1998).

119. Rose, *The Politics of Life*, 57.

120. Agamben, *Homo Sacer*, 142.

121. Paul W. Kahn, *Sacred Violence: Torture, Terror, and Sovereignty* (Ann Arbor: University of Michigan Press, 2008), 14.

122. Irving Louis Horowitz, *Taking Lives: Genocide and State Power*, 3rd ed. (New Brunswick, NJ: Transaction Books, 1980), xi–xiii.

123. Michel Foucault, "The Political Technology of Individuals," in *Power: Essential Works of Foucault, 1954–1984*, vol. 3, edited by Paul Rabinow (New York: New Press, 2000), 403–17; at 409.

Chapter Two

The State Must Own Death: Germany

Marianne, a seventeen-year-old girl from Czechoslovakia, arrived at Auschwitz sometime in early 1943. Her reception was hellish, chaotic. She recalls,

> We arrived at night. . . . Because you arrived at night, you saw miles of lights—and the fire from the . . . crematoria. And then screaming and the whistles and the "Out, out!" . . . , and the uniformed men and the SS with the dogs, and the striped prisoners—we, of course, at that time, didn't know who they were—and they said, "Throw everything out. Line up—immediately."[1]

And amid the confusion, the screaming, the horrific smell was the selection process. Marianne continues:

> They separated you and then lined up everybody in fives, . . . and there were two men standing. . . . On one side, was the doctor, one was [Joseph] Mengele, . . . and on the other side was the . . . *Arbeitsführer*, which was the . . . man in charge of the work *Kommando*. And it was . . . "You go, you go by truck. You walk, you go by truck." . . . A pattern pretty soon developed that you could see—under fourteen about and over thirty-five were assigned to the trucks. And not until we actually marched into the camps did you know exactly where the trucks had gone, . . . and this was done . . . very fast, very efficient.[2]

Those directed to the trucks were sent to the gas chambers. They would be forced to remove all valuables, undress, and under threat of whip or gun, sent to what appeared to be a shower. The doors would be closed and in a matter of minutes all would be dead; their corpses hurriedly removed, destined either for massive trenches for burning, or to the furnaces for cremation.

Figure 2.1. A Jewish woman walks toward the gas chambers with three young children after going through the selection process on the ramp at Auschwitz-Birkenau, May 1944. *Source:* Yad Vashem, courtesy of the United States Holocaust Memorial Museum

Figure 2.2. Jewish women and children who have been selected for the gas chambers at Auschwitz-Birkenau, May 1944. *Source:* Yad Vashem, courtesy of the United States Holocaust Memorial Museum

For many survivors of the Nazi Holocaust, nothing is more emblematic of the calculated valuation of life and death as the selection ramp. For it was here, most clearly, that one person assumed the position of God: to determine who shall live and who shall die. And perhaps most salient is that these life and death decisions were made within parameters established by the state. And while not legal as we understand that concept, the decisions to live or die were not entirely illegal under the Nazi regime. Such was the liminality surrounding the valuation of life under Nazi rule. Consider, for example, that on July 19, 1942, Heinrich Himmler provided a "legal" basis for the mass killings that were already under way:

> I herewith order that the resettlement of the entire Jewish population of the General Government [Nazi-occupied Poland] be carried out and completed by December 31, 1942. From December 31, 1942, no persons of Jewish origin may remain within the General Government, unless they are in the collection camps in Warsaw, Cracow, Czestochowa, Radom, and Lublin. . . . These measures are required with a view to the necessary ethnic division of races and peoples for the New Order in Europe, and also in the interests of the security and cleanliness of the German Reich and its sphere of interest. Every breach of this regulation spells a danger to quiet and order in the entire German sphere of interest, a point of application for the resistance movement and a source of moral and physical pestilence. For all these reasons a total cleansing is necessary and therefore to be carried out.[3]

Nearly one-third of the world's 19 million Jews—approximately 6 million people—perished in the Holocaust. Geographically, the figures are numbing, as numerous European countries lost at least 70 percent of their Jewish population. In Poland, for example, approximately 2.9 million Jews died, constituting 90 percent of the prewar Jewish population; the death of 65,000 Jews in Yugoslavia, likewise, represented nearly 90 percent of that country's prewar total.[4] Coupled with the death of much of Europe's Jewish population are the millions of others: Roma and Sinti (Gypsies), homosexuals, "asocials," "handicapped," Jehovah's Witnesses, and so on. In total, approximately 15 million persons died as a result of Nazi genocidal policies.[5] And this number still does not include the millions of other people who died on the battlefields of World War II.

The Holocaust was composed of *state-sanctioned* violent practices that stemmed from a basic geographic imagination: to construct a pure living space for one population through the elimination of another population. Moreover, the planners of the Holocaust, as Koonz explains, "followed a coherent set of severe ethical maxims derived from broad philosophical concepts."[6] These include a mixture of racial and geopolitical theories that, combined, were used to determine who might live and who must die. As Peter

Fritzsche writes, the Nazis "are frightening because they expanded notions of what is politically and morally possible in the modern world."[7] He explains that for the Nazi leaders, "worldviews could bring a world into view" and, as such, their geographical imaginations "posed the question of life or death, national survival or annihilation, in the most radical terms." In short, the elimination of the Jews, the Sinti, the handicapped, and so on "indicate that the greater German empire the Nazis set out to create rested on the intention to violently remake 'lands and people' into 'spaces and races.'"[8] Although directed primarily against "the Jew," the Nazi state in effect assumed the right to eliminate those populations that posed—in their utopian worldview—a threat to their existence: "those without a *right* to exist, who *ought* not to exist, and who were therefore *obligated* not to exist."[9] The Holocaust proceeded on the paradoxical belief that in order to promote life, life itself must be sacrificed. However, this necessitated a deliberate calculation of life, a determination of lives worth living and lives unworthy of life. Roberto Esposito drives this point home: "From time immemorial racial persecutions have been based on the presupposition that the death of some strengthens the life of others."[10] In short, the Holocaust was born from a utopian fantasy whereby life sprang from death, and communal inclusion was derived from bodily exclusion.

Beyond the nearly incomprehensible magnitude of the killing—the human toll—lies an even more insidious facet of the state-sanctioned industrial killing of the Holocaust: the widespread participation of millions of Germans. Apart from a small handful of decision makers, including Adolf Hitler, Heinrich Himmler, Reinhard Heydrich, Joseph Goebbels, and Adolf Eichmann, the systematic and bureaucratic slaughter resulted directly from the participation of an army of scientists and other intellectuals, all of whom crafted a policy and constructed a physical infrastructure designed to efficiently exterminate an entire population. German racial theorists, physicians, lawyers: all of these so-called intellectuals "appealed not so much to malevolence as to ideas of health, hygiene, and progress in their campaign to elicit compliance with policies that might otherwise have been seen as cruel and violent."[11] Thus remains the simple, yet chilling, observation that "the Holocaust could not have happened without the participation of many other Germans—soldiers, civil servants, doctors, lawyers, industrialists, clerks, railway workers, even clergymen—who were not ideologically committed Nazis of long standing yet who helped locate Jews, identify them, imprison them, and kill them without protests or attempts at evasion."[12] The purpose of this chapter, therefore, is to consider state-sponsored violence, to work through ideas of sovereignty and the spatiality of life and death. In subsequent chapters on the Great Famine in China and genocide in Cambodia, I extend my arguments to consider other valuations of life and death.

EUGENICS AND RACIAL COMPETITION

"Racial ideologies and theories," as Burleigh and Wippermann write, "were not an exclusive German discovery." However, as these two authors conclude, the Nazi state "became the first state in world history whose dogma and practice was racism."[13] In short, the Nazi state was constituted by a form of *state racism* that gave legitimacy to the taking of life in order to make life. The sociospatial practices of "emigration, deportation, ghettoization, Aryanization, enforced euthanasia, genocidal sterilization, calculated starvation, socially engineered disease, excessive labor, mass shooting, mobile gassing, mass gassing, mass burning and death marches,"[14] suggest that as the Nazi state developed, power—embodied in the figures of Hitler, Himmler, Heydrich, and myriad others—was viewed as the right to kill in order to live. This was made possible, as Michel Foucault suggests, because of state racism, for it was at this point that racism intervened to separate bodies into populations: populations categorized as "us" or "them," "good" or "evil," "friend" or "enemy." Foucault states bluntly: "It is primarily a way of introducing a break into the domain of life that is under power's control: the break between what must live and what must die."[15] It is a way of separating types of bodies, of constructing different populations that are evaluated in such a way as to justify death over life. It is a biopolitical practice predicated on "rational calculations of instrumental reason."[16] Foucault explains further:

> The appearance within the biological continuum of the human race of races, the distinction among races, the hierarchy of races, the fact that certain races are described as good and that others, in contrast, are described as inferior: all this is a way of fragmenting the field of the biological that power controls. It is a way of separating out the groups that exist within a population. It is, in short, a way of establishing a biological type caesura within a population that appears to be a biological domain.[17]

The racial regrouping of bodies into populations and subpopulations is a markedly spatial process. It is an activity of dividing and separating bodies for political and economic purposes. However, there is an additional *sovereign* component to these sociospatial separations, one that is immediately connected to questions of state rule. Foucault writes: "The fact that the other dies does not mean simply that I live in the sense that his [*sic*] death guarantees my safety; the death of the other, the death of the bad race, of the inferior race (or the degenerate, or the abnormal) is something that will make life in general healthier: healthier and purer."[18]

Nazi scientists did not invent race; nor did they substantially alter existent theories of race. Rather, they drew upon the accumulated "knowledge" of Western ideas and concepts of race. Contemporary Western theories of "race" originated concurrently with the emergence of nationalism during the eighteenth and nineteenth centuries. Situated at the apex of European imperialism, these were the years of exploration and observation, classification and conquest, as scientists sought to make sense—and order—of the world they "discovered."[19] Out of the scientific discourse emerged, ostensibly, objective criteria for understanding the geographical distribution and diffusion of races. Classifications were constructed according to observed—and selected— physical traits and were employed to explain perceived cultural traits. Gradually, these racial ideas would merge with nascent geopolitical conceptions of statecraft—including that of the Nazi state.

Racial classifications were ordered hierarchically and geographically. Northern and western European races and civilizations were held as cultural yardsticks against which all other races and societies were measured. The most common standard included a tripartite division of the "white" race: the Nordics, Alpines, and Mediterraneans. The tall and fair-skinned Nordics (e.g., those of British or German descent) were supposed to represent the highest echelons of human evolution. Lower down the evolutionary ladder were found the shorter and somewhat darker Alpines (e.g., people from Poland and Hungary), followed by the much shorter and darker Mediterraneans (e.g., people from Greece and Italy). In contrast to the "white" race were the various races of the "yellow," "brown," and "red" civilizations. "Blacks" or "Negroes" were at the bottom of the racial hierarchy. As Sarah Danielsson concludes, the "devastating assumption that humans differed significantly and could be divided into distinct 'groups' was taken for granted, the questions only revolved around *how* different and in *what ways*."[20]

During the late nineteenth century, the scientific paradigm underwent a philosophical shift from a study of racial comparison to that of racial competition. Evolutionary theory, proposed and elaborated independently by Charles Darwin and Alfred Russel Wallace, forewarned the notion of a struggle for existence among different species. In particular, the publication of Darwin's *On the Origin of Species* (1859) undermined fixed biological categories of race and suggested that humans evolved through various natural laws of competition. This reconfiguration of human evolution—and, by extension, of racial classification—was adopted by Herbert Spencer, who forthrightly claimed that purposeful cruelty was nature's method for biological progress.[21] The shift in racial research was ominous: no longer was the study of race classification merely an exercise. Increasingly, races were thought to be engaged in a monumental competition for survival.

As the twentieth century approached, concern grew among many scholars, politicians, and pundits that contemporary society—through charity and welfare programs—were "tampering" with the laws of nature; that unfit people were being allowed to survive and reproduce; that "lives not worth living" were posing an unbearable financial burden on state economies. These political valuations of life identified a grievous and imminent social problem and demanded an immediate solution. Progress in social science—and a concomitant political will—was needed to understand the implications of human interference and, if possible, provide viable solutions. What was required, in practical terms, was a biopolitical control over the three fundamental processes of population change: migration, fertility, and mortality.

Biopower, as a set of practices and techniques for achieving the subjugation of bodies and the regulation of populations, saw its apex in the form of eugenics. Originating from the work of Francis Galton on biological inheritance during the 1860s, the eugenics movement effectively combined growing racist and nationalist sentiments of the late nineteenth and early twentieth centuries. As Saul Dubow writes, eugenics coincided with the rising intensity of imperialist feelings, which also helped augment nationalist fervor and provide a convenient rationale for the colonial subjugation of non-European bodies—including Germany's imperial adventures throughout the African continent.[22]

Apart from the Darwinian model of evolution, eugenics incorporated other intellectual strands, such as August Weismann's theory of the "continuity of germ plasm." According to Weismann, there were theoretical and experimental grounds for thinking that only a portion of each cell—its *germ plasm*—carried hereditary material. Independent from the rest of the cell, germ plasm could be inherited from one generation to the next—and without alteration from outside influences. As such, Weismann's theory was a direct challenge to another dominant theory of evolution, that being Jean Baptiste Lamarck's theory of the inheritance of acquired characteristics. Following Lamarck's theory, environmental influence and experience could be transmuted generationally. If, for example, a cat were to lose its tail in an accident, we would expect that the offspring of that feline would have truncated tails. Weismann's theory, conversely, held that external influences had no effect on subsequent generations. Stepan explains that if Weismann's ideas about the continuity of the germ plasm were correct, then the effects of education and improved surroundings would not be assimilated genetically over successive generations.[23]

Given the diversity of competing ideas and concepts circulating among the scholarly community, it should come as no surprise that there was no singular theory of eugenics; rather, the term itself referred to a collection of assorted beliefs and prescriptions that were manifest differently in different places.[24] In general, however, eugenical discourses retained a central tenet in that a distinction between superior and inferior people, or "races," could be identified and that science held the tools for implementing racial improvement.[25] Moreover, the essential "traits and qualities" of races were held to be biologically bound, and that any social or spatial proximity between races portended the potential for racial or societal degeneration. Writing in the United States, John Commons, for example, asserted that "race differences are established in the very blood and physical constitution. They are most difficult to eradicate, and they yield only to the slow processes of the centuries."[26] When combined with the emergence of geopolitics (discussed later), a eugenical discourse intimated the potential for race wars.

As a paradigm, the doctrine of eugenics sought not only to identify differences in human populations, but to control—to manage—populations according to some *a priori* concept of superior/inferior peoples. As explained by the eugenicist L. H. M. Baker, "When we have ascertained . . . the qualities we want to preserve and the characteristics we desire to eliminate[,] we must be courageous in the application of our remedy."[27] From the United States to Germany, from Japan to Mexico, an array of segregation policies, immigration legislation, relocation schemes, antimiscegenation laws, and sterilization and euthanasia programs emerged as tangible instruments in the practice of statecraft. The eugenics movement epitomized state intervention in the spatiality of life and death.

By the early twentieth century, eugenics societies began to be formed in order to pursue scientific investigations and to promote new policies and legislation.[28] Consequently, it bears repeating that "eugenic solutions to real, or perceived, social ills had widespread currency long before the Nazis came to power—and by no means exclusively in Germany."[29]

Eugenically based solutions can broadly be classified as positive or negative. Positive eugenics includes policies to increase the racial contribution of populations deemed the most desirable. In their widely used textbook *Applied Eugenics*, for example, Paul Popenoe and Roswell Hill Johnson argued that the birthrate of the American stock was too low and that, therefore, the most desirable seed stock was dying out and being supplanted by immigrants.[30] The basis of negative eugenics—including such practices as sterilization and euthanasia—was to reduce the biological contribution of the least desirable populations. Writing in 1914 in the United States, the sociologist Edward Ross warned of "conquest made by child-bearing" of blacks and immigrants.[31] Similar pronouncements were made in Germany, such as by Otto Kankeleit, the German sterilization expert who demanded that the sterilization of "inferior" women should become a priority.[32]

Throughout the early decades of the twentieth century, therefore, eugenicists and their followers retained discursive influence over the preservation of race and state. Fueled by Darwinian concepts of sexual selection and evolution, competing concepts of genetics, Malthusian perspectives on overpopulation, and a basic layer of environmental determinism, widespread fears of racial and societal degeneration (in both popular and academic venues) were kindled in the late nineteenth and early twentieth centuries. These fears were augmented by a growing awareness—again, both in popular and academic circles—of demographic changes and economic hardships. Out of this scientific milieu emerged a dominant perspective of humans and their geographical environment, namely, that human activities were understood to follow natural laws, dictated by the geographic environment or innate "racial" qualities.[33]

THE GEOPOLITICS OF RACE WARS

Murray Edelman suggests that "perception involves categorization" and that "categorizations give meaning both to what is observed and to what is assumed."[34] When these categorizations involve territories, they may be considered *geographies*. Situating the emergence of modern "geography" as an academic field within imperial systems, Gearóid Ó Tuathail views geography as a verb: to *geo-graph*. Closely associated with politics, "geo-graphy" to Ó Tuathail signifies an attempt to organize and control territories and populations. The political representation of geography, in other words, assumes a regulatory function. This is the domain of geopolitics.[35]

Geopolitics, as a field of study and a legitimating feature of statecraft, emerged alongside eugenics. And similar to many eugenicists, early geopoliticians often were profoundly influenced by the writings of Charles Darwin, Herbert Spencer, and other proponents of evolutionary thought, as concepts such as competition, survival, and health figured prominently in their writings. As Nancy Ordover writes, the "eugenics project revolved around imagining the nation: what it was (now threatened) and what it might be (with and without government and medical intervention)." For Ordover, it "was the sort of creative visualization that demanded both historical revisionism and ominous prophecy."[36] The articulations of geopoliticians provided just such *geographical* imaginations.

The geopoliticians of the late nineteenth and early twentieth century were divided into two main schools: geostrategists and proponents of the organic state theory.[37] Geostrategists, in general, emphasized the importance of immutable geographical positions, such as oceans, waterways, and landmasses. Their scale of analysis was mostly at the global level. Key theorists included the American Alfred Thayer Mahan and the British Halford Mackinder. The political and ideological differences of these theorists stemmed in part from their personal backgrounds and their different readings of history, geography, and technology. The naval historian Mahan, for example, focused on the lessons of the British Empire and stressed commercial expansion through sea power. Mackinder, conversely, drew insights from the civilizations of Central Asia and touted the importance of land power.

In contrast to the geostrategists, geopoliticians such as Friedrich Ratzel and Rudolf Kjellén viewed states as analogous to living organisms. Ratzel, drawing on his formal training in the natural sciences as well as evolutionary theory, based his political geography on the concept of *lebensraum* or "living space." To Ratzel, every "form of life needs space in order to come into existence, and yet more space to establish and pass on its characteristics."[38] States operated in like manner; confronted with expanding populations and dwindling resources, expansion was imperative for the survival of a state. It was Ratzel who championed the idea that "between the movement of life which is unceasing, and the extent of the earth which is unalterable, there is a struggle. And from this struggle the war for space was born."[39]

During the early twentieth century, Ratzel's theories were expounded by Kjellén, a Swedish geographer and Germanophile. In his 1916 book *The State as Form of Life*, for example, Kjellén takes Ratzel's metaphor of the state *as* organism quite literally. He understands the state "to be a 'living form' . . . to the extent that it is furnished with instincts and natural drives."[40] As Esposito writes, this constituted a radical transformation of the idea of the state, "according to which the state is no longer a subject of law born from a voluntary contract but a whole that is integrated by men and which behaves as a single individual both spiritual and corporeal."[41] This would have grave implications in the coming years, for it was Kjellén who argued that it was the struggle for space that comprised the most important elements of human history; that extermination was a necessary product of gaining vital *lebensraum*; that the will to self-preservation and the will to power were the driving forces behind the struggle to the death between different peoples.[42]

According to Danielsson, what is particularly important about the concept of *lebensraum* is that it "had an immediate and wide impact on geographers, political scientists and anthropologists."[43] Indeed, even before the Nazi state used *lebensraum* as a rationale for their eastern European expansion and the Holocaust, these geographic concepts formed the basis of Germany's colonization and genocide in Southwest Africa.[44] Of significance was the presumed connection between territory (space) and populations as manifest in the theoretical imbrications of geopolitics and eugenics. As Bassin explains, the vision of the international arena as the scene of an ongoing struggle grew increasingly popular: a struggle in which national interests were necessarily in conflict and in which one nation's gain could mean nothing but another's loss and possible decline—if not death.[45]

The confluence of geopolitics and eugenics produced a worldview—and one that was not limited to Germany—in which racial proximity and territorial expansion would lead to racial and societal degeneration. Utilizing metaphors of contamination and degeneration, purity and security, the eugenically based geopolitics provided keystones for the construction and preservation of racial states. Scientists, as well as politicians, purported to identify essential and unchanging differences in races; the command of discourse was used to legitimate racist and nationalist policies whereby individual liberties were discarded in the name of racial and national preservation. Populations were the object of political calculations; individuals were to be sacrificed for the greater good of the state. Charles Davenport, a noted eugenicist, effectively captures the schism between individual rights and state security: "Where the life of the state is threatened, extreme measures may and must be taken." He continues: "Society must protect itself; as it claims the right to deprive the murderer of his life so also it may annihilate the hideous serpent of hopelessly vicious protoplasm. Here is where appropriate legislation will aid in eugenics and in creating a healthier, saner society in the future."[46] A decade later, Adolf Hitler would echo these sentiments in his book *Mein Kampf.*

To ensure survival of the state, many policy makers (both in Germany and beyond) advocated the retention of a healthy, vigorous population; this to them tended to imply a racially homogenous population. A geopolitically informed eugenics policy, therefore, demanded the identification of "inferior" and "degenerate" bodies, those whose lives threatened the security of race and state—a precursor to our contemporary discussions of "biosecurity." Such was the politically informed calculation of life and death that marked Nazi Germany.

Viewed from this perspective, the targeting of Jews and other "undesirable" peoples assumes a deeper geographical dimension. Rather than a continuation of earlier anti-Semitic attitudes, the violence that became the Holocaust was part and parcel of a much broader "demographic reorganization" that entailed three interrelated components: the "ethnic cleansing of Jews from the Third Reich, of Poles from the Third Reich, and the repatriation of ethnic Germans from abroad."[47] Browning explains further that this "grandiose program of demographic engineering" was "based on racial principles that would involve the uprooting of millions of people. These policies were fully consonant with Hitler's underlying ideological assumptions: a need for *lebensraum* in the east justified by a Social-Darwinist racism, a contempt for the Slavic populations of Eastern Europe, and a determination to rid the expanding German Reich of Jews."[48] As Bergen makes clear, "in their choices of target groups the Nazis reflected and built upon prejudices that were familiar in many parts of Europe. Hitler and the Nazis did not invent anti-Semitism nor were they the first to attack Sinti and Roma (Gypsies) or people considered handicapped."[49] Likewise, one cannot forget that many of the theories, policies, and practices that underlie the Holocaust were *not* unique to the Nazi state. As Nikolas Rose explains, "In the education of German citizens in the Third Reich, in eugenic education campaigns in the United States, Britain, and many European countries, making social citizens involved instructing those citizens in the care of their bodies—from school meals to toothbrush use, inculcation of the habits of cleanliness and domesticity, especially in women and mothers, state regulation of the purity of food, interventions into the workplace in the name of health and safety, instructing those contemplating marriage and procreation on the choice of marriage patterns, family allowances, and much else."[50]

LIFE, DEATH, AND THE NAZI STATE

"We all accept that killing is in general wrong," Jeff McMahan writes, "but virtually all of us also recognize certain exceptions—that is, we concede that there can be instances in which killing is permissible."[51] Killing in self-defense, for example, is often held as an acceptable reason for taking another's life; likewise, killing in defense of one's country—such as in war—is also considered to be just. Other forms of killing are justified, although often subject to intense debate. Abortion, physician-assisted euthanasia, and capital punishment have been, or remain, legal in many countries throughout the world.

In chapter 1 the concept of sovereignty and the right to life was discussed. It was observed that in many different political-philosophical traditions a common feature is that the sovereign retains the right over life and death. In the classical understanding of sovereignty, this right assumed the form of "taking life" or "letting live." More recently—and largely following the emergence of the modern nation-state, this sovereign right was transformed as a right to "foster life" or to "disallow it to the point of death." However, as readily apparent in debates over abortion, euthanasia, and capital punishment, the modern sovereign state continued to retain the right to determine when, if ever, it is acceptable to "take life": to kill.

This presumption of life and death epitomizes the Nazi state. According to Giorgio Agamben, in Germany there emerged an unprecedented situation whereby the absolute authority to "make live" intersected with an equally absolute authority to "make die."[52] Such was the sovereign basis that attempted to legitimate and justify the Nazi's pending race war.

From its beginnings in 1933, the nascent Nazi state epitomized the twin technologies of power that constitute *biopower*: one technique was disciplinary and centered on the body, while a second brought together the mass effects of a population and attempted to shape demographic processes toward an ultimate political objective. The two forms of power were thus complementary: one directed toward specific bodies and the other toward populations.

Nazi biopower, however, did not emerge fully formed in 1933; rather, the two techniques of biopower—anatomo-politics and biopolitics—developed over time, as Hitler and his associates experimented—literally—with the lives of the German people. As Bergen explains, Nazi members "struck in dramatic, decisive ways, but they always tested the public response to each move before proceeding further."[53] In typical fashion, Nazi leaders would introduce a policy, a practice, or piece of legislation, evaluate the response of the citizenry, and then move on. Thus, although the Holocaust—the ultimate annihilation of the entire Jewish population—may have been a long-term dream of Hitler, Himmler, and others, it was, in the short term, neither attainable nor necessarily conceivable in 1933. Hitler was well aware that to put the machinery in motion that would construct the Nazi utopia—to achieve the needed *lebensraum* that would be *Judenfrei* ("Jew free")—it was necessary to first *socially kill* the Jews and other undesirable bodies.

Here, the conceptual work of Orlando Patterson is especially informative. In his seminal work on slavery, Patterson develops the concept of "social death." He answers his own question, that if a slave "had no social existence outside of his [*sic*] master, then what was he?" with the response that the slave was a "socially dead person." By this Patterson explains that, in part, the slave "is desocialized and depersonalized." However, this is incomplete. Instead, Patterson draws attention to two ways in which social death was enacted. On the one hand, there occurred an "intrusive mode of representing social death" whereby "the slave was ritually incorporated as the permanent enemy on the inside."[54] Accordingly, within the Nazis' geographical imagination, Jews were not, and could not be, part of Germany. Jews, in essence, were permanently outsiders, foreign to the body politic. This concept, in fact, conforms readily with Agamben's idea of *homo sacer*, in that Jews could only be "included" in the German state through their "exclusion."

On the other hand, and in contrast to the intrusive conception of death, is an "extrusive" representation. Here, according to Patterson, "the dominant image of the slave was that of an insider who had fallen, one who ceased to belong and had been expelled from normal participation in the community because of a failure to meet certain minimal legal or socioeconomic norms of behavior." Indeed, Patterson explains that "the destitute," similar to slaves, "were likewise included in this group, for while they perhaps had committed no overt crime their failure to survive on their own was taken as a sign of innate incompetence and of divine disfavor."[55] From this perspective, Jews (and other "asocials") were considered part of the German state, but only as "fallen" bodies. For members and supporters of the Nazi Party, these unwanted bodies were *in* Germany but not *of* Germany. As David Clarke and his coauthors write, the "abhorrence of the Holocaust lies in the fact that to be classified in this way was to be judged as a body without a right to exist, and as a life that was obligated to cease to exist."[56] As Esposito writes, Nazism "treated the German people as an organic body that needed a radical cure, which consisted in the violent removal of a part that was already considered spiritually dead."[57]

The social death of Germany's unwanted populations began in earnest following the appointment of Hitler as chancellor on January 30, 1933. At first, Hitler and the Nazi Party targeted suspected political opponents—notably members of the Communist Party—through extralegal practices. Members of opposition groups were harassed, humiliated, beaten, and murdered; others were imprisoned in makeshift detention centers and newly constructed concentration camps, such as Dachau, which opened in March of that year. The Nazi Party created also its own political opportunities—seemingly to operate within legal boundaries. On February 27, 1933, for example, Nazi members burned down the Reichstag—the German parliament building. Hitler, however, blamed the arson on German communists and was therefore able to "justify" his continued repression on his political opponents. More important, Hitler used the Reichstag fire as justification for the passage of the Enabling Law of March 23, 1933. With this law, Hitler was able to put through any measure without approval from parliament. Democracy, in effect, ceased to exist in Germany.[58]

In 1933 the first of a series of laws was passed that would codify specifically the social death of the Jewish population. The Law for the Restoration of the Professional Civil Service, passed on April 7, 1933, sanctioned the dismissal of "non-Aryans" and others considered to be politically undesirable from public service; subsequent laws passed on April 22 and April 25 removed or restricted Jews from a range of professions.[59] In effect, these laws were designed to strip Jews of their German citizenship, thus legally and spiritually placing them outside human and divine law: they were, following Agamben, becoming *bare life*. They were simultaneously included and excluded from the newly emerging German community.

Hitler and his followers called for a new start, a reawakened Germany, which would produce a racial or national community (*Volksgemeinschaft*).[60] The notion of "community" is thus of particular importance. The modern theoretical and empirical study of communities is traced to the German sociologist Ferdinand Tönnies and his 1887 publication *Gemeinschaft und Gesellschaft*. Tönnies identified what he believed to be a *fin-de-siècle* shift, wherein traditional communities (*Gemeinschaften*), built around the ideas of kinship, cooperative action, and attachment to place, were being threatened, or supplanted, by more impersonal and artificial societies (*gesellschaften*). It is significant that the Nazi state focused on the tradition-bound *Volkgemein- schaft*. As Koonz concludes, "Nazi rule relied not only on repression but also on an appeal to communal ideals of civic improvement."[61]

"Community" is an explicitly *geographical* concept. In ordinary usage, the term generally refers to the people with whom one identifies in a specific locale.[62] It is a concept that evokes "commonality"; an inclusive social group that shares particular values, beliefs, and morals. Community likewise implies a sense of security and solidarity: a feeling of togetherness that is often infused with a sense of purpose. For many, there is an implicit assumption that a "lack" of community is a bad thing.[63]

Communities must be both defined and defended. As Stuart Aitken writes, if we assume that "community," with its vague and generally nurturing meanings, is usually something that people desire, then it must be something people do not yet have (or fear losing) in the way that they want.[64] The promotion of community is thus inescapable from both a discussion (and enforcement) of inclusion and exclusion. As David Sibley writes, "Who is felt to belong and not to belong contributes in an important way to the shaping of social space."[65] Amartya Sen captures this dual nature of community. He notes, on the one hand, that a "sense of identity can make an important contribution to the strength and warmth of our relations with others, such as neighbors, or members of the same community, or fellow citizens, or followers of the same religion. Our focus on particular identities can enrich our bonds and make us do many things for each other and can help to take us beyond our self- centered lives." However, on the other hand, Sen concedes that "a sense of identity can firmly exclude many people even as it warmly embraces others. The well-integrated community in which residents instinctively do absolutely wonderful things for each other with great immediacy and solidarity can be the very same community in which bricks are thrown through the windows of immigrants who move into the region from elsewhere."[66] Or the very same people who imprison, sterilize, castrate, and murder noncommunity members.

When we ask who has the authority to define and classify populations, and to include or exclude people on those grounds, we are properly talking about sovereignty and our right to space. And usually, when we ask who has the authority to determine who might marry, or reproduce, or be killed, we are talking about state sovereignty. However, as Koonz explains, "in modern societies experts create assumptions about which people belong within the community of shared moral obligation." In Nazi Germany it was the doc- tors and physicians, professors and lawyers, who "provided the knowledge

. . . about which humans deserved moral consideration."[67] Jews, Sinti and Roma, homosexuals, criminals: all found themselves under the medical and legal gaze of state-sanctioned authorities who held in their hand the right to participate in the German community.

Throughout the Third Reich, numerous intellectuals—not all of whom were members of the Nazi Party—produced volumes of scholarly treatises that would provide the academic rationale and legal standing for a range of discriminatory policies and practices. On July 14, 1933, for example, Hitler enacted the Law for the Prevention of Hereditarily Diseased Offspring. Also known as the Sterilization Law, this law permitted the compulsory sterilization of persons suffering from allegedly hereditary illnesses. Consequently, if in the opinion of a state-sanctioned "genetic health court" the patient was determined to suffer from any one of a list of illnesses or conditions, he or she would be forcibly sterilized. The practice of sterilization within the Nazi state is discussed in greater detail later; for now, it is important to note that the practice of involuntary sterilization "required the participation of many people: lawyers who drafted the legislation; medical and social workers who reported people to be sterilized to the authorities; bureaucrats who handled the paperwork; doctors, nurses, and aides who performed the procedures."[68]

The Nazi state took a step closer to its goal of a biologically pure *lebensraum* on September 15, 1935, with the passage of the Nuremburg Laws. The first of these laws, the Law for the Protection of German Blood and Honor, criminalized marriage or sexual relations between "Aryans" and "non-Aryans"; forbade Jews from flying the German flag; and prohibited Jews from employing German gentile women under age forty-five in their households. In total, this law served to both disenfranchise German Jews and to ensure a social (and sexual) separation from Aryans and all others. The second law, the Reich Citizenship Law, provided a legal foundation for the category of "Jew." Under the law, being Jewish was based on the religious practice of one's grandparents. Accordingly, if three or more of one's grandparents were considered to be Jewish, then that individual was declared a Jew. However, if one had two grandparents who were Jewish, the person in question was classified as *Mischlinge*, or "mixed blood." Based on the Reich Citizenship Law, Jews were denied German citizenship; those classified as *Mischlinge* occupied a more liminal position. Significantly, the law did not attempt to define "Aryan," nor, for that matter, did the German scholarly community ever define Aryan.[69]

Whereas the Nuremburg Laws of September 1935 were directed primarily against a Jewish "contamination" of the German *Volk*, other laws reveal that for the Nazi state, biological threats were found in non-Jews as well. Both the Law against Dangerous Habitual Criminals (passed on November 24, 1933)

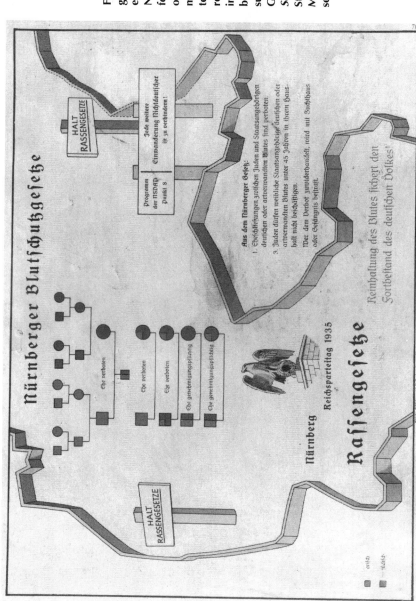

Figure 2.3. Eugenics poster entitled "The Nuremberg Law for the Protection of Blood and German Honor." The text at the bottom reads, "Maintaining the purity of blood ensures the survival of the German people." *Source:* United States Holocaust Memorial Museum

and the October 18, 1935, Law for the Protection of the Genetic Health of the German People were enacted to maintain the "health" and vitality of the German *Volk*. The earlier law provided for the detention and compulsory castration of selected criminals, while the latter law, also known as the Marital Health Law, constituted an attempt to "protect" German blood and honor from defilement. In particular, this latter law required couples to submit to medical examinations prior to marriage; the law also made possession of a "certificate of fitness to marry" mandatory for all prospective marriage partners.[70] Marriages were prohibited if either individual was found to be suffering from venereal disease, feeblemindedness, epilepsy, or any other "genetic" condition specified in the 1933 Sterilization Law. Exceptions to the marriage ban were permitted only on the condition that the couple was sterilized.[71] Taken together, these two laws codified both positive and negative eugenics, and served to foreshadow the more extreme practices that were to follow.

The pending genocide in Nazi German was a spatial practice, designed and implemented to purify the country—the community—of undesirable elements. The enactment of a series of laws throughout the 1930s, such as the Marital Law and the Sterilization Law, all addressed a deepening rationalization of who was to be included—and who was to be excluded—from the German *Volk*. Communal participation was predicated upon a valuation

Figure 2.4. Propaganda slide featuring a disabled infant. The caption reads, ". . . because God cannot want the sick and ailing to reproduce." *Source:* United States Holocaust Memorial Museum

of populations: a calculation of life and death. Who was to be included or excluded; who would be granted life or death; who would be allowed to contribute, quite literally, to the reproduction of the community?

Koonz argues that "the Nazi state removed entire categories of people from the German's moral map" based on four guiding principles.[72] The first principle, heavily informed by geopolitical thought, held that "the life of a *Volk* is like that of an organism, marked by stages of birth, growth, expansion, decline, and death."[73] Related, a second principle was that "every community develops the values appropriate to its nature and to the environment within which it evolved."[74] The Nazi state's third principle, again following the geopolitical theories of Ratzel and Kjellén, maintained that "outright aggression against 'undesirable' populations living in conquered lands" was justified.[75] And last, a fourth principle—related to the sovereign right to take life—upheld the right of a government to enact "state-sponsored ethnic cleansing."[76] In short, and following Esposito, we must acknowledge that "it isn't enough to conclude . . . that the limits between healing and killing [were] eliminated in the biomedical vision of Nazism. Instead we need to conceptualize them as two sides of the same project that makes one the necessary condition of the other: it is only by killing as many people as possible that one could heal those who represented the true Germany."[77]

THE NAZI STATE OF VIOLENCE

From 1933 onwards, the Nazi Party worked to construct their racially pure community. They were enjoined by legions of professionals—racial anthropologists, biologists, and hygienicists, economists, geographers, historians, sociologists, statisticians, social workers, doctors and physicians, nurses and midwives—not all of whom were active members of the Nazi agenda. But for the most part, the acquiescence of these professionals reflects that they were largely in support of the broader goals, the racial dreams, forwarded by Hitler and his associates.[78] Michael Burleigh and Wolfgang Wippermann are blunt in their condemnation of these professionals:

> Imprisoned in the little worlds of their laboratories, libraries, and institutes, and captive to professional hierarchies, careerism, the quest for "fame," and petty professional animosities, many of these professionals failed to see [or refused to see] the barbaric and inhuman goals which their research served. . . . Contrary to the notion that Nazism somehow corrupted and distorted the temples of learning—which of course it did—one could argue that a corrupt and inherently distorted science lent Nazism a specifically "academic" and "scientific" character.[79]

The Nazi's program of creating a "pure" *lebensraum* was from the outset future oriented. Nowhere is this more clearly revealed than in the eugenically informed practice of sterilization. Here, the Nazi state set out to prevent the birth of people whose lives, many years later, would be viewed as an obstacle to German progress.

Sterilization had long been debated—and advocated—both in Germany and beyond. Indeed, a number of highly influential eugenicists argued for the prevention of the "genetically defective," possibly by state-sanctioned sterilization. Charles Davenport, whom we encountered earlier, surmised that if the state could take a person's life (i.e., capital punishment), then the state was also within its sovereign right to deny the right of reproduction.[80] According to Ordover, the first scientific proposal for sterilization published in the United States was written in 1887 by the superintendent of the Cincinnati Sanitarium, targeting prisoners; six years later, a Texas doctor—who supported castration of homosexual men—proposed that sterilization be used for the "purpose of race improvement."[81] It was not until 1907, however, that the first state in the United States legalized compulsory sterilization. In that year the Indiana legislature made legal the sterilization of confirmed "criminals, idiots, rapists, and imbeciles." In less than two decades, twenty-four states had passed involuntary sterilization laws, which were used to sterilize mainly poor (and often African American) inmates of institutions for the feebleminded; and these laws were acted upon. By the end of World War II, American physicians had sterilized upwards of 70,000 people—most without their consent.[82] And throughout Denmark, Sweden, Norway, Finland, Iceland, and Germany sterilization likewise was viewed as a viable medical practice to prevent societal degeneration. As a means of effecting a calculated management of life—the prevention of lives that posed a social and financial burden to society—sterilization was for many scientists and physicians a low-cost, scientifically informed solution.[83] Ironically, though, sterilization was illegal in Germany until 1933.[84]

But who was to be prevented from reproducing? In the United States, for example, eugenicists targeted the ambiguous categories of hereditary criminals, imbeciles, morons, and the feebleminded—these latter being "scientific" measures of intelligence. It was widely presumed that both mental abilities and criminal proclivities were genetic, and thus passed from generation to generation. Sterilization, as such, would put an end to the endless, downward spiral of degeneration. In the US Supreme Court's 1927 decision of *Buck v. Bell*, Justice Oliver Wendell Holmes wrote:

> We have seen more than once that the public welfare may call upon the best citizens for their lives. It would be strange if it could not call upon those who

already sap the strength of the State for these lesser sacrifices . . . in order to prevent our being swamped with incompetence.[85]

In this passage, Holmes echoes the concerns expressed in Germany, namely, the question of why states could demand sacrifice of its soldiers, yet ask no less of those persons who contributed nothing to society: the imbeciles, morons, idiots, and feebleminded. Holmes continued: "It is better for all the world if, instead of waiting to execute degenerate offspring for crime, or to let them starve for their imbecility, society can prevent those who are manifestly unfit from continuing their kind."[86] For Holmes, and the 8–1 majority of the US Supreme Court, it was affirmed that the state not only retained the sovereign right to kill (i.e., execution) or to let die, the state also had the right to prevent life through the sterilization of those deemed genetically unfit.

Similar discussions were held in Germany throughout the late nineteenth and early twentieth century. Of particular importance was Adolf Jost's 1895 essay *Das Recht auf den Tod* ("The Right to Die"). Foreshadowing the American Supreme Court ruling, Jost—who introduced the concept of *negative Lebenswert* ("life without value")—argued that the right of life and death belonged to the state and, as such, that the state was well within its sovereign right to authorize the direct medical killing of people. Jost's argument conformed with the broader geopolitical imaginings of Ratzel and other political theorists, namely, that where "the health of the political body as a whole is at stake, a life that doesn't conform to those interests must be available for termination."[87] In effect, as Lifton writes, Jost's position was that "the state must own death—must kill—in order to keep the social organism alive and well."[88]

It was not, however, until the 1920 publication of *Die Freigabe der Vernichtung lebensunwerten Lebens* ("The Permission to Destroy Life Unworthy of Life") that negative eugenical practices, including sterilization, gained wide traction in Germany. Co-written by Alfred Hoche, professor of medicine, and Rudolf Binding, professor of law, this publication effectively synthesized the biological and juridical justifications for the taking of "lives unworthy of life." Hoche and Binding asserted that the right to live must be earned and justified; those not capable of human feeling, described as "ballast lives" or "empty human husks," could have no sense of value of life; theirs is a life not worth living; destruction is not only tolerable, but humane.[89] They argued that the destruction of lives unworthy of life was in fact a pragmatic act, in that these lives were essentially already "dead." In Hoche's section, for example, he wrote that putting people to death "is *not* to be equated with other types of killing . . . but [is] an *allowable, useful act*."[90] To this end, Hoche supported his argument with a calculated valuation of life. He claimed that "20–30 idiots with an average life expectancy of fifty represented 'a mas-

Figure 2.5. Propaganda slide featuring two doctors working at an unidentified asylum for the mentally ill. The caption reads, "Life only as a burden." *Source:* United States Holocaust Memorial Museum

sive capital in the form of foodstuffs, clothing and heating, which is being subtracted from the national product for entirely unproductive purposes.'"[91]

Both Hoche and Binding, moreover, make the connection between the state's sovereign right to ask soldiers to sacrifice for their country and the practice of keeping (and paying for) the feebleminded to be housed in asylums. They explained: "If one thinks of a battlefield covered with thousands of dead youth and contrasts this with our institutions for the feebleminded with their solicitude for their living patients—then one would be deeply shocked by the glaring disjunction between the sacrifice of the most valuable possession of humanity on one side and on the other the greatest care of beings who are not only worthless but even manifest negative value."[92] As Koenisberg concludes from this passage, the message for the Nazi Party was clear: "If the state was willing to sacrifice the lives of its soldiers, why should so many resources be expended to keep mental patients alive?"[93]

Curiously, for Binding and Hoche, the state-authorized taking of "lives unworthy of life" was *natural*: it was already inscribed into those lives. As Esposito finds, it "is as if the right/obligation to die, rather than falling from on high in a sovereign decision on the body of citizens, springs from their own vital makeup. In order to be accepted, death must not appear as the negation but rather as the natural outcome of certain conditions of life."[94] In other words, the political termination of life sprang not from lawyers or doctors but rather from life itself. It was only natural that some lives were indeed unworthy of life—their "lives," in fact, were *already* devoid of life. As Esposito explains, the perception of these bodies as "empty human husks" or "human ballast" "has precisely the objective of demonstrating that in their case death does not come from outside, because from the beginning it is part of those lives—or, more precisely, of these *existences* because that is the term that follows from the subtraction of life from itself. A life inhabited by death is simply flesh, an existence without life."[95]

Their deaths, however, were not to be sacrificed; lives unworthy of life were not synonymous with those of soldiers who gloriously died for their country. Rather, these lifeless lives were beyond law. The mentally ill, hereditary asocials, homosexuals, racial Others, Jews were all *homo sacer*, lives that could be killed with impunity, their deaths registered as neither homicide nor sacrifice. They existed simply to be eliminated; death had no meaning beyond a means to an end. For Esposito, what the Nazis wanted to kill in the Jew and in other lives not worthy of life was not, ironically, life, but the presence in life of death: a life that was already dead because it was marked hereditarily by an original and irremediable deformation.[96]

From the first sterilizations sanctioned by the Nazi state, through to the death camps, the quest for *lebensraum* was predicated upon the elimination of life unworthy of life. Consequently, rather than medical killing being subsumed to war, the war itself was subsumed to a vast biomedical and geographical vision of which eugenical solutions were the primary weapons. Or, to put the matter another way, the deepest impulses behind the war had to do with the sequence of sterilization, selective euthanasia, and mass extermination.[97]

Ironically, throughout the years of Nazi rule, it remained illegal to kill a human being, except by soldiers in battle or criminals legally convicted by a court of law.[98] As Henry Friedlander explains, "the pre-Nazi penal code was never abolished, and its articles 211 and 212 prohibiting the intentional killing of a human being remained in full force and effect." Friedlander continues that Hitler refused to promulgate a law that would authorize killings, on the grounds that such a law would make public the overall scope of the program.[99]

This is significant. The many physicians and civil servants who participated in mass killings wanted *legal* authorization; they wanted to be sure that they would not be criminally liable. Hitler, for his part, refused to issue a direct order to that effect. He did however provide in October 1939—predated to the onset of war on September 1, 1939—a letter that served as a legal basis for all subsequent killing operations. Although the letter did not have the force of law, it provided the veneer of state sanction.[100]

THE PRACTICE OF DEATH

From the outset, the Nazi regime made sterilization the first application of the biomedical imagination to the issue of collective life or death. The first Sterilization Law, of July 14, 1933, according to Lifton, "set the tone for the regime's medicalized approach to 'life unworthy of life.'"[101] In practice, the law provided for the compulsory sterilization of the hereditarily ill, defined as anyone who suffered from congenital feeblemindedness, schizophrenia, manic depression, hereditary epilepsy, Huntington's chorea, hereditary blindness, hereditary deafness, or serious physical deformities. Furthermore, those individuals afflicted with "chronic alcoholism" were also to be involuntarily sterilized. The law itself was the product of a series of recommendations put forward by the Committee of Experts for Population and Racial Policy, established under the direction of Reich Minister of the Interior Wilhelm Frick.[102]

The research of Robert Proctor, Michael Burleigh, and Robert Lifton, in particular, vividly reconstruct the functioning of the Nazis' sterilization program. Within one year of the law, 181 "genetic health courts" (*Erbgesundheitsgerichte*) and "appellate genetic health courts" (*Erbgesundheitsobergerichte*) were established throughout Germany. The health courts, usually attached to local civil courts, were composed of three members, two physicians and a district judge. After 1935 these courts were also responsible for rendering decisions on marriages.[103]

All physicians were legally required to report to health officers anyone they encountered in their practices who suffered from any of the prescribed conditions. Those who refused were fined. In practice, the law created more *practical* problems than it did *moral* problems. Having interviewed many doctors associated with the Nazi state, Robert Lifton concluded that the majority of doctors approved of the sterilization laws; they believed that the laws were consistent with prevailing medical and genetic knowledge concerning the prevention of hereditary defects.[104]

Of greater concern was how to maintain the traditional privacy of the doctor-patient relation, given that the doctors' records were to be submitted to the health courts. After numerous deliberations, it was decided that physicians would report cases of genetic illnesses much as they were required to report births, deaths, and infectious or venereal diseases.[105]

A variety of medical procedures were utilized in the attempt to secure a more hygienic future. Most commonly, men underwent vasectomies while

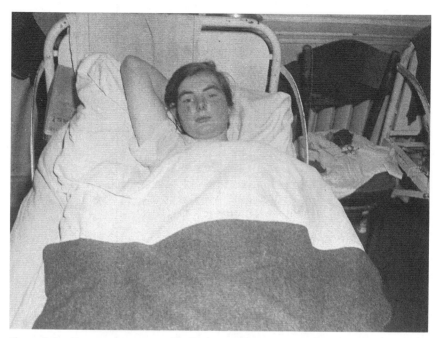

Figure 2.6. Twenty-three-year-old Elizabeth Killiam, the mother of twins, was sterilized in Weilberg before being transferred to the Hadamar Institute. *Source:* Rosanne Bass Fulton, courtesy of the United States Holocaust Memorial Museum

women endured tubal ligation. The procedure for men was relatively straightforward, and therefore time and cost efficient. The operation for women, however, was decidedly more complex; a hospital stay of eight to fifteen days was required, compared to the five-to-ten-minute procedure for men. Accordingly, physicians experimented with procedures that would allow for a more rapid sterilization on an "out-patient" basis. Such "techniques," many of which were developed in the concentration camps (discussed later), included the injection of caustic substances (e.g., carbon dioxide) into women's cervixes and exposing men's penises and scrota to radiation and chemicals.[106]

It remains unclear as to how many men and women were sterilized during the Nazi period; most scholars provide a range between 300,000 and 350,000 involuntary sterilizations, with a few writers suggesting upwards of 400,000.[107] Whatever the true number, it is clear that the sterilization program entailed the participation of a large number of people. Proctor explains that "doctors competed to fulfill sterilization quotas; sterilization research and engineering rapidly became one of the largest medical industries. Medical

supply companies made a substantial amount of money designing sterilization equipment. Medical students wrote at least 183 doctoral theses exploring
the criteria, methods, and consequences of sterilization."[108] To this we can
add the judges, lawyers, paralegals, and other administrative workers who
effectively managed the entire program.

From Sterilization to Euthanasia

By 1938 the Nazis' program to eliminate lives not worthy of life underwent
a transformation, as planners and physicians shifted their focus from future
to present threats. No longer content with eliminating the next generation of
unwanted persons, practitioners within the Nazi state set about eliminating
those already born. At this point, the Nazi state initiated a process that no
other eugenically informed state had ever attempted: to "progress" from mass
sterilization to mass "euthanasia."

The term belies its horrific perversion by the Nazi state. Derived from
two Greek words, *eu*, meaning "good," and *thanatos*, meaning "death," euthanasia is simply defined as "good death." Conceptually, euthanasia (often
equated with suicide) occupied the attention of Western philosophers since
at least the time of ancient Greece. Both Socrates and Plato, for example,
regarded suffering as a result of a painful disease to be a sufficient reason
for ending one's life. Likewise, following the example set forth by Socrates,
who himself committed suicide, there may be moral, or justifiable political
reasons, for taking one's life. That said, other Greek philosophers, notably
Aristotle, Pythagoras, and Epicures, condemned euthanasia as a practice.[109]

Christian belief also prohibited euthanasia. St. Augustine, for example,
argued that suicide was in opposition to the sixth commandment, and that
life and suffering were divinely ordained by God; to take one's life was to
go against God, for only God had the right to decide when a person would
die. However, associated with the Protestant Reformation and the subsequent
separate of Church and State, other philosophers and political theorists began to associate the "right" of life or death with that of the sovereign. Both
Thomas More and David Hume argued that euthanasia was acceptable for
terminally ill patients.[110]

Throughout Europe and North America, into the twentieth century, the
subject of euthanasia was widely debated and increasingly garnered more
support. In 1935 the British Voluntary Euthanasia Society was founded,
followed three years later by the establishment of the Euthanasia Society of
America.[111] The discussion of euthanasia in Germany, both prior to and during the Nazi regime, was hardly unique.

It is important to establish, conceptually, the parameters of euthanasia.
On the one hand, it is possible to distinguish euthanasia based on the actions

involved. Killing an individual for his or her own good is referred to as *active* euthanasia, while "letting" a person die is considered "passive" euthanasia. Physician-assisted suicide, where lethal injections are administered by a presiding doctor, would be considered a form of active euthanasia; the removal of a patient from life-support systems would be considered passive euthanasia. On the other hand, euthanasia may be classified by various forms of "consent." *Voluntary* euthanasia refers to a situation where a person freely and autonomously requests or consents to be killed or allowed to die for his or her own good; *nonvoluntary* euthanasia, conversely, exists when it is not possible for an individual who is killed or allowed to die either to give or to withhold consent. In this latter situation, which would include "congenitally severely cognitively impaired" persons, some other "responsible" individual makes the final decision. Lastly, euthanasia is said to be *involuntary* when an individual who is competent to give or to withhold consent is killed or allowed to die either contrary to his or her expressed will or when his or her consent has not been sought.[112]

Significantly, within Germany, Jost and other writers began to erase the distinction between voluntary euthanasia and the involuntary killing of "life unworthy of life"; Jost believed that the "incurably ill" and "mentally defective" could be killed regardless of whether those slated to die were in a position to articulate their wishes.[113] In fact, many writers began to advocate that even parents should not have the right to decide. Only the sovereign state, emotionally detached and with a concern for the broader *Volk*, should have that right.

In *Mein Kampf*, Hitler raised the question—Why do the best die while the worst survive?—in moral terms.[114] This question, as the preceding section makes clear, addresses broader moral, political, and economic calculations of lives worth living. This valuation of life, however, was not directed at any given individual, but rather at populations—entire groupings of people: the mentally ill, the physically deformed, homosexuals, criminals, the Sinti and Roma, the Jews. Ultimately, the decision to kill, or to let die, was predicated on whether entire populations—as defined by the Nazi state—were to be included or excluded from the *Volk*. The decision was asked first of children.

As Lifton disturbingly writes, "It seemed easier—perhaps more 'natural' and at least less 'unnatural'—to begin with the very young: first, newborns; then, children up to three and four; then older ones."[115] The impetus arrived in the form of a letter. In the fall of 1938 Hitler received a petition from a German father asking that his child—born blind, physically deformed, and mentally ill—be granted a "mercy death." Hitler subsequently ordered his own personal physician, Karl Brandt, to evaluate the child and to consult with the attending doctors. Brandt reported to Hitler that the doctors agreed that "there was no justification for keeping [such a child] alive"; after the war,

Brandt elaborated that "it was pointed out [in 1938] that in maternity wards in some circumstances it [was] quite natural for the doctors themselves to perform euthanasia in such [cases] without anything further being said."[116] It was decided that the child be "allowed" to die.

For Hitler, this single death, this isolated act of "letting" die, served both as inspiration and opportunity. By May 1939, Hitler asked Brandt "to appoint an advisory committee to prepare for the killing of deformed and retarded children."[117] Three months later, on August 18, 1939—fourteen days before the invasion of Poland—the Committee for the Scientific Treatment of Severe, Genetically Determined Illnesses delivered a secret report to all state governments. According to the report, all midwives or doctors who delivered any infant with congenital deformities were required to register that child with local health authorities. Such deformities included "idiocy or Mongolism (especially if associated with blindness or deafness); microcephaly or hydrocephaly of a severe or progressive nature; deformities of any kind, especially missing limbs, malformation of the head, or spina bifida; or crippling deformities such as spastics." Furthermore, doctors were ordered to report any child in their care up to the age of three and suffering from any of the specified deformities to local health offices.[118]

District medical officers were subsequently responsible for ensuring the "accuracy" of the reports, which were then submitted to three medical experts who, without ever directly evaluating the children, would render judgment. Those selected for "mercy killing" were immediately ordered into one of twenty-eight institutions that were (ultimately) established throughout Germany, Austria, and Poland.[119] In the end, more than 5,000 children were killed in this first phase of the German euthanasia program.[120]

The methods of killing varied based in part on the attitudes of the doctors. It was not uncommon for children to be administered sedatives, such as Luminal tablets that were dissolved in tea or water. These sedatives would be given repeatedly, often in increasing doses. After two or three days the child would slip into a coma and die. Conversely, some children would be injected with morphine or other medicines, thereby bringing about a more rapid death. And still other children were denied food or left in rooms without any heat; either way, the children were allowed to "let die" whether from starvation or exposure.[121] These latter methods, according to some physicians, were considered to be the simplest and most cost-effective.

The T-4 Program

The decision to murder "defective" infants and children was followed by a decision to eliminate those adult lives also considered not worthy of living:

the mentally ill and physically handicapped. SS officer Christian Wirth, who would work at a number of "euthanasia" sites before becoming commandant of the death camp Belzec and then inspector of the Belzec, Sobibor, and Treblinka death camps, captured the overall Nazi attitude: "Mental patients are a burden upon Germany and we only want healthy people. Mental patients are a burden upon the state."[122] According to the head of the Reich Chancellery, Hans Heinrich Lammers, Hitler is likewise reported to have explained that

> he regarded it as right that the worthless lives of seriously ill mental patients should be got rid of. He took as examples the severe mental illnesses in which the patients could only be kept lying on sand or sawdust, because they perpetually dirtied themselves, cases in which these patients put their own excrement in their mouths as if it were food, and things similar. Continuing on from that, he said that he thought it right that the worthless lives of such creatures should be ended, and that this would result in certain savings in terms of hospitals, doctors and nursing staff.[123]

In October 1939—but backdated to September 1 to coincide with the start of the war—Hitler issued a brief decree. It reads: "Reich Leader [Philip] Bouhler and Dr. Brandt are charged with the responsibility for expanding the authority of physicians, to be designated by name, to the end that patients considered incurable according to the best available human judgment [*menschlichem Ermessen*] of their state of health, can be granted a mercy death [*Gnadentod*]."[124] Hitler's decree did not have the force of law; it did not provide legal sanction.[125] However, this was consistent with the overall conduct of the Nazi Party.

The "adult euthanasia" program "was a carefully planned and covertly executed operation, with precisely defined objectives."[126] Burleigh elaborates that "those responsible believed in the necessity of what they were doing. Mentally and physically disabled people were killed to save money and resources, or to create physical space for ethnic German repatriates."[127] To accomplish these objectives, a camouflage organization called the Reich Work Group of Sanatoriums and Nursing Homes was established. Operating from the Berlin Chancellery at its Tiergarten 4 address, the program became known as Aktion T-4.[128]

The T-4 program was, at its core, a "rational," calculated management of life. According to Proctor, the "original intent of those who planned the euthanasia operation was to scale it according to the formula 1,000:10:5:1—that is, for every 1,000 Germans, 10 needed some form of psychiatric care; 5 of these required continuous care; and among these, 1 should be destroyed."[129] Given that Germany's population was approximately 65 to 70 million, this empirical calculation called for the killing of 65,000 to 70,000 people.

That those slated for death were to be "lives unworthy of life" was immediately clear. But how were specific individuals to be chosen? In part, the planners followed the child-murder operation. Burleigh and Wippermann explain that the "Reich Ministry of the Interior dispatched forms to all asylums and clinics to be completed for each patient. These supplied details about the latter's race, state of health, and capacity to work. The completed forms were then centrally processed and duplicated, and sent in batches of 150 to expert 'assessors.'"[130] For services rendered these "assessors," all of which were physicians, received a monetary payment. In their "evaluation"—which did not include a in-person visit with the patient in question—the assessors would mark the forms with a red + if the individual was to die, or a blue − if the individual was to allowed to let live. For those cases requiring further evaluation, a higher panel of medical experts would evaluate the file. Euthanasia applications began in late 1939 and the first executions of adult mental patients began in early 1940.[131]

It must be stressed that both the child-murder operation and Aktion T-4 were conducted in the utmost secrecy. Although those who planned and participated in the operations were supportive of the overall objectives—to provide a healthy *Volk*—the parents of the intended victims did not wish to see their loved ones killed. It was the parents, the grandparents, the aunts and uncles, and brothers and sisters who entrusted their loved ones to the doctors and physicians; it was the family members who placed their relatives into hospitals and asylums—to receive medical care and attention, not death. Consequently, planners for these programs developed elaborate schemes to convince the surviving family members that their loved one died a natural death.

Secrecy was not the only problem. Unlike the relatively straightforward practices developed to kill children, administrators of Aktion T-4 were unsure as to how to effectively and efficiently go about the mass murder of potentially tens of thousands of patients. Consequently, officials conducted various tests to determine the best techniques of mass elimination. In early 1940 an experiment was conducted at a psychiatric hospital at Brandenburg, near Berlin, whereupon a group of patients was to be killed in a newly devised gas chamber (under the supervision of Christian Wirth), while another group was to be killed by injection. The patients who were injected "died very slowly" whereas those who were gassed (with carbon monoxide) died rapidly. Moreover, the gas chamber could accommodate mass killings very easily. Afterward Brandt testified: "This is just one example of [what happens] when major advances in medical history are being made. There are cases of an operation being looked on at first with contempt, but then later on one learned it and carried it out. Here the task required by state authority was added to the medical conception of this problem, and it was necessary to find with good conscience a basic method that could do justice to both of these elements."[132]

Six main killing centers were established in former mental hospitals or prisons—Hartheim, Sonnenstein, Grafeneck, Bernburg, Brandenburg, and Hadamar; all were equipped with specially outfitted gas chambers (disguised as showers) and crematoria to burn the bodies.[133] Burleigh describes the sequence of operations. Once patients were selected for death, they would be taken from their clinic/asylum and bused to a "holding" or "transit" center located near one of the actual killing centers. This intermediary step "enabled T-4 [administrators] to minimize any inconvenience to themselves by precisely staggering transport, killing and cremation." From the transit center, patients were then driven to one of the aforementioned six sites. On arrival, the patients were marched into the facility where they were ordered to undress and then given military overcoats to cover themselves. An administrator checked each person's identity and a doctor provided a cursory examination. The purpose of this screening was to establish a plausible cause of "natural" death, which would be reported to the patient's family. Patients were next weighed, photographed, and given a cardboard number so that they might ostensibly retrieve their clothes after showering. Some patients would be marked with chalk letters.[134]

Having been undressed and evaluated, the patients were hurriedly moved to the gas chambers; depending on the size of the chamber, perhaps sixty patients could be killed at one time. A doctor nearby would turn a valve, releasing the gas. Burleigh continues: "So far from it being a 'gentle death,' the victims experienced extreme terror, as well as the symptoms of carbon monoxide poisoning. After an hour, all was quiet. The ventilators extracted residual fumes, and the 'burners' or 'disinfectors' . . . moved in to disentangle the bodies. Those who had been marked beforehand as being of potential scientific interest [marked with chalk] were separated out and taken to a nearby autopsy room. Their brains would go to the university clinics in Frankfurt or Würzburg. All of the corpses were trundled along the corridor and, after gold teeth had been wrenched out, burned in the . . . crematoria. The ashes were either discarded or distributed among urns. . . . Bones were crushed in a mill or with mallets on wooden tables."[135]

The program operated with factory-like precision; within twenty-four hours of arrival, patients would be processed, killed, and cremated. All that was left was for the administrators to send prewritten condolence letters to the victims' families, expressing concern for the untimely, though natural, death that had befallen their loved one. And for the doctors and physicians—for those who were directly responsible for the decision to kill or to let live? It was a job; but it was also a mission. As Proctor writes, "Doctors were never

Figure 2.7. The corpse of a woman lies in a coffin at the Hadamar Institute where she was put to death as part of the Operation T-4 euthanasia program. *Source:* Rosanne Bass Fulton, courtesy of the United States Holocaust Memorial Museum

Figure 2.8.　Photograph of a survivor at the Hadamar Institute taken after liberation in 1945. She was committed to the facility five years earlier after being judged insane for having a Jewish boyfriend. *Source:* Rosanne Bass Fulton, courtesy of the United States Holocaust Memorial Museum

ordered to murder psychiatric patients and handicapped children. They were *empowered* to do so, and fulfilled their task without protest, often on their own initiative."[136] Burleigh concurs, noting that "there is no recorded instance of any person being punished for refusing to carry out such killings."[137]

Most killing centers were only in operation for a matter of months. Grafeneck, for example, operated between January and December 1940; and Bernburg was in operation only between February and September of that year. In this short amount of time, however, the program met its intended objective. By August 1941, when the program officially came to an end, 70,273 people had been killed—a figure that matches well with the calculated number of expected deaths that had been established by the T-4 planners.[138]

The official stoppage of Aktion T-4 did not bring an end to the mass killing; rather, it ushered in a period of even greater killings. Burleigh writes that the "euthanasia" program *officially* ended "because its team of practiced murderers was needed to carry out the infinitely vaster enormity in the East that the regime's leaders were actively considering."[139] It was at this point that the Nazis' will to space expanded beyond Germany and into the occupied territories of Poland and the Soviet Union.

The Will to Space

For years Nazi officials had planned on a massive demographic reshuffling of Europe. This would entail the removal of millions of undesirable populations—namely the Jews—and the resettlement of eight million ethnic Germans into the newly acquired territories. Bergen writes that the "more Hitler's empire extended its reach and consolidated its hold on subject lands and peoples, the more its forces sought to destroy those it deemed enemies— Jews above all, but also Slavs and others defined as unwanted inferiors."[140] She explains that German planners developed in detail their goal of achieving *lebensraum* in a memorandum called "General Plan East"—a document that in effect provided the coordinates for a new population geography. Specifically, the plan "called for Germans and ethnic Germans to settle vast areas of Eastern Europe, where they would produce food and babies for the 'Aryan master' race. To make such settlements possible, the plan demanded expulsion of those currently living there. Those tens of millions of people, most of them Slavs, were to be forced into less desirable areas, allowed to die of starvation and disease, or turned into slaves for the German empire. According to General Plan East, Jews were to disappear altogether."[141]

Previously, Nazi leaders had contemplated simply the removal of Jews through mass emigration. In fact, many of the early concentration camps (see later) had been used as techniques of terror, to induce Jews (and other undesirable populations) to leave "voluntarily." Other plans called for the forcible relocation of Jews to ever more distant lands, such as the island of Madagascar off the coast of Africa. Most of these plans, however, came to naught. Conversely, the successes of the euthanasia programs seemed to offer a more efficient and "final" solution. There is not, of course, a direct order from Hitler authorizing the "Final Solution." As Rees explains, the Nazi state did not function that way. Rather, "Hitler set the mood, had the vision, thought the unthinkable; others devised the policy. Hitler wanted it; others did it."[142]

At some point during the fall of 1941, Hitler, Himmler, and Heydrich, in particular, "were aware that the ultimate goal or vision of Nazi Jewish policy was now the systematic destruction and no longer the decimation and expulsion of all European Jews."[143] Browning elaborates that "in mid-July 1941 Hitler instigated Himmler and Heydrich to undertake what amounted to a 'feasibility study' for the mass murder of European Jews, and in early October 1941 he shared with Himmler and Heydrich his approval of their proposal to deport the Jews of Europe to killing centers in the east."[144] Arad concurs, echoing that the "period between the outbreak of war and the German invasion of the Soviet Union on June 22, 1941, was the transition from a policy of forced emigration to one of physical annihilation."[145]

Browning is not without his detractors; others, for example, suggest that the transition to mass murder occurred later, possibly following the Wannsee Conference—originally scheduled for December 12, 1941, but eventually held on January 20, 1942. However, it does appear that the decision to eliminate the Jews occurred prior to the conference; this latter assemblage served to clarify that the extermination would take place. As Rees explains, the Nazis had decided to exterminate a population. The question was, how?

Early acts of mass murder were "hidden" by the "fog of war," most notably through the deployment of special death squads, known as *Einsatzgruppen* ("special action groups"). Einsatz units, composed of between 500 and 1,000 men, were task forces that had been established to follow the German army in its conquests of Austria, Czechoslovakia, and now, Poland. According to Rhodes, "Einsatzgruppen secured occupied territories in advance of civilian administrators. They confiscated weapons and gathered incriminating documents, tracked down and arrested people the SS considered politically unreliable—and systematically murdered the occupied country's political, educational, religious, and intellectual leadership."[146] According to Bergen, during the summer of 1941 the objectives—the purpose—of these groups underwent a profound transformation. Increasingly, these units began to interpret their primary job as slaughter of all Jews, including women, children, and old people as well as the murder of Sinti, Roma, and inmates of mental hospitals.[147]

Far from delivering a "good death," the methods of these death squads were brutal, primitive, and calculated to instill terror throughout the occupied lands. Bergen details that "many of the actions of the mobile killing units more or less followed the same pattern. First they rounded up the Jews in a given area using various ruses to deceive them and relying on local collaborators for denunciations. The Germans ordered large pits dug in some convenient area—a local cemetery, nearby forest, or easily accessible field. Often they forced the prisoners themselves to dig what would be their own graves. At gunpoint they made the victims undress. Then they shot them by groups directly into the graves. In this manner the mobile killing units and their accomplices killed around a million people."[148]

During August of 1941, Himmler visited the east and witnessed firsthand the mass shootings; he was disturbed by what he saw and sympathized with the "psychological burden" placed on the killers—hence the need to find less traumatic ways to kill large numbers of people.[149] Moreover—and arguably more important—the mass shooting of victims was *too slow*. Despite

Figure 2.9. Mass execution and grave site associated with the liquidation of the Mizocz ghetto, Poland, October 14, 1942. *Source:* Instytut Pamieci Narodowej, courtesy of the United States Holocaust Memorial Museum

Figure 2.10. Corpses of prisoners exhumed from a mass grave in the vicinity of Hirzenhain, Germany, May 1945. Approximately 250 slave laborers, 200 of whom were women, were interned in a factory near this site. *Source:* Nathan Weil, courtesy of the United States Holocaust Memorial Museum

the massive numbers of Jews, Slavs, and others killed, the actions of the *Einsatzgruppen* were not sufficient to effect the Nazis' vision of a cleansed eastern Europe.[150] And it was in part for this reason that Aktion T-4 came to an abrupt end, for the industrialized process of quickly killing large numbers of people at home was needed in the east.

Nazi planners understood that it was more efficient to bring victims to a centralized site of killing; and, as Friedlander writes, it was only logical that these places be modeled on the Aktion T-4 centers. On December 8, 1941, the first explicit "death camp" was established, located at Chelmno in Poland. Although considered a "stationary killing center," it used mobile gas vans to carry out the murders.[151] According to Yitzhak Arad, by the middle of 1942 about thirty gas vans had been produced by a private car manufacturer to facilitate the killings.[152]

The first "true" death camps began operation in the spring and summer of 1942. Three camps, Belzec, Sobibor, and Treblinka, were equipped with stationary gas chambers in which a diesel motor propelled carbon monoxide into the chambers.[153] Two other concentration camps, Auschwitz and Majdanek, were refitted to serve as killing centers. Just as the sterilization and euthanasia programs did not simply materialize out of thin air, neither did the death camps. Indeed, the death camps can be viewed as the end point of a series of policies, programs, and practices that all revolved around the Nazis' will to *lebensraum.*

Nazi officials had long used concentration camps as a means of instilling terror, inducing mass emigration, and, for certain individuals, "letting die" through starvation, exposure, or exhaustion. Initially, Nazi officials utilized former prisons, jails, and workhouses as sites of detention. In 1933 most Germans had not backed the Nazi Party and political opposition remained a legitimate threat. Accordingly, apart from public beatings and murders, Nazi leaders rounded up thousands of opposition members—or those suspected of supporting the opposition.

The "camp system" began to take shape with the opening of the first official concentration camp, Dachau (near Munich), in March 1933. And over the next decade, Dachau would emerge as the prototypical camp—complete

Figure 2.11. Newly arrived prisoners, with shaven heads, stand at attention during a roll call in the Buchenwald concentration camp, November 10, 1938. These prisoners were among the approximately 10,000 German Jews who were arrested during the November 9–10 Kristallnacht pogrom and sent to Buchenwald. *Source:* Robert A. Schmuhl, courtesy of the United States Holocaust Memorial Museum

with guard towers and barbed wire. Other camps would open as necessity demanded. Following political purges or pogroms directed against Jews, Nazi leaders found themselves in need of more camps. However, as the Nazi Party solidified its position—but before it embarked on its massive program of mass murder—the camp system contracted in size. By summer of 1935 there were just five concentration camps still in existence, holding fewer than 4,000 prisoners.[154]

With the approach of war in 1938 and 1939 conditions changed. On the one hand, the concentration camps were increasingly used to facilitate the still-hoped-for mass emigration of Jews. The mass detentions that followed the terror of *Kristallnacht* is a case in point. Between November 9 and 10, 1938, Nazi followers fomented a pogrom that resulted in the destruction of hundreds of synagogues, businesses, and other Jewish institutions. About 100 Jews were killed, while another 25,000 to 30,000 were arrested and interned at the Buchenwald, Dachau, and Sachsenhausen concentration camps.[155] Tellingly, many of these detainees were released within a few days; Nazi officials hoped that targets of the recent terror would decide to "voluntarily" leave Germany.[156]

On the other hand, German industrialists and Nazi planners worried of impending labor shortages. Himmler and other Nazi officials recognized that concentrated groups of prisoners—the epitome of "docile labor"—could easily be exploited as slave laborers. Consequently, new camps were founded, such as those at Flossenburg and Mauthausen, in order to contribute to the war effort.[157]

By 1942, as the Germany military claimed new territories in Poland and the western Soviet Union, Nazi officials were confronted with additional problems, namely, the millions of Jews, Slavs, and other unwanted populations acquired through the aforementioned General Plan East. Apart from mass shootings conducted by the *Einsatzgruppen*, Nazi officials utilized another form of concentration: the establishment of ghettos.

Beginning in late 1939, German officials demanded that all Jews residing in the Polish territories be rounded up and detained in designed urban areas (ghettos). People were forced, at gunpoint, to flee their homes; entire villages and towns were violently cleared of Jews and other undesirables. Those who resisted, or were unable to leave, were shot on sight.

The Jewish ghettos, such as in Lodz and Lublin, emerged as stopgap measures: to concentrate the Jews until their fate could be determined by the Nazi state. In the interim, Nazi officials established "Jewish Councils" to administer aspects of the daily lives of the ghettos. It was placed on the shoulders of these councils to distribute resources, organize social life, and determine who was fit for work and—later—who would be selected to die.[158]

Conditions in the ghettos were horrendous: severely overcrowded, unsanitary, often lacking in water or sewage. Bergen explains that in 1940 the ghetto at Lodz held 230,000 people in only 30,000 apartments; only about 725 of those lodgings had running water. The inhabitants, moreover, were forced to produce materials for the German war effort. By 1943 ghetto workshops were producing uniforms, boots, underwear, and bed linen for the German military; other goods of metal, wood, leather, fur, and paper, as well as electrical and telecommunication devices, were also made.[159] Under these conditions, thousands died, from starvation, exposure, disease. Although in principle these ghettos were not designed to effect mass murder, Nazi officials were more than content to "let die" those Jews imprisoned.

The cessation of Aktion T-4 and the promotion of General Plan East would cast the ghettos in a new light. Increasingly, these ghettos no longer were viewed as "prisons" or sources of exploitable slave labor; instead, the ghettos became "transit points," as the inhabitants were to be shipped for elimination at the newly built death camps.[160] The death camps operated on assembly-line procedures with large gas chambers that required limited manpower and would kill in relative secrecy.[161] At some camps, upwards of *10,000* people could be killed on a single day.

Figure 2.12. Corpses at the recently liberated Dachau concentration camp, c. May 1945. *Source:* Henry Pitt, courtesy of the United States Holocaust Memorial Museum

McKale provides a glimpse into the scale and scope of the infrastructure that served to facilitate the massive population movements: "Between the fall of 1941 and spring of 1945, more than 260 deportation trains hauled German, Austrian, and Protectorate Jews to the ghettos and extermination camps in Poland and Russia. Other trains took victims to Theresienstadt, the ghetto-transit camp near Prague for many elderly Germans and decorated Jewish war veterans. Further, approximately 450 trains ran from Western and Southern Europe to the death camps: a minimum of 147 trains from Hungary, 87 from Holland, 76 from France, 63 from Slovakia, 27 from Belgium, 23 from Greece, 11 from Italy, 7 from Bulgaria, and 6 from Croatia."[162]

And thus we return to the beginning—to the selection ramps on which a seventeen-year-old girl from Czechoslovakia found herself one terrifying evening in 1943: a night, like so many others during the Holocaust, where individuals, operating on behalf of a sovereign state, calculated the worth of life or the savings of death.

CONCLUSIONS

Esposito affirms that "to assume the will of power as the fundamental vital impulse means affirming at the same time that life has a constitutively political dimension and that politics has no other object than the maintenance and expansion of life."[163] The progression from forced emigration, segregation, and sterilization to mass murder was neither inevitable nor linear. Rather, these various "demographic" practices were developed, and functioned, in parallel. All constituted attempts to achieve the overarching goal of the Nazi state: to provide living space (*lebensraum*) for its population (*Volk*). Early forms of segregation (i.e., imprisonment in concentration camps and ghettos) were implemented to accomplish two main goals: first, to separate, and thus to maintain the purity between different "populations"; and second, to encourage mass emigration from German territory. Sterilization was promoted to prevent future generations from existing, while euthanasia was to eliminate the present populations. By 1941–1942, these parallel practices merged, as unwanted populations were concentrated for the explicit purpose of mass elimination: the development of death camps. Even these camps, however, resulted from processes of trial and error (e.g., mass shootings), as Nazi leaders sought to develop the most effective and efficient means of mass murder.

In the next chapter we see a vastly different valuation of life, whereby the state sacrifices its own citizenry in the pursuit of a utopian fantasy. In Maoist China, the value of a person was determined not so much on presumed racial traits, but rather on productive ability. In so doing, the state allowed upwards of 40 million people to die.

NOTES

1. Quoted in Robert J. Lifton, *The Nazi Doctors: Medical Killing and the Psychology of Genocide* (New York: Basic Books, 1986), 163.

2. Lifton, *The Nazi Doctors*, 163–64.

3. Quoted in Yitzhak Arad, *Belzec, Sobibor, Treblinka: The Operation Reinhard Death Camps* (Bloomington: Indiana University Press, 1999), 47.

4. There is a remarkable literature on the Holocaust. Among the many starting points, see Raul Hilberg, *The Destruction of the European Jews* (Chicago: Quadrangle, 1961); George L. Mosse, *Toward the Final Solution: A History of European Racism* (New York: Harper & Row, 1978); Robert N. Proctor, *Racial Hygiene: Medicine under the Nazis* (Cambridge, MA: Harvard University Press, 1988); Zygmunt Bauman, *Modernity and the Holocaust* (Cambridge, MA: Polity Press, 1989); Michael Burleigh and Wolfgang Wippermann, *The Racial State: Germany 1933–1945* (Cambridge: Cambridge University Press, 1991); Henry Friedlander, *The Origins of Nazi Genocide: From Euthanasia to the Final Solution* (Chapel Hill: University of North Carolina Press, 1995); Omer Bartov (ed.), *The Holocaust: Origins, Implementations, Aftermath* (New York: Routledge, 2000); Donald M. McKale, *Hitler's Shadow War: The Holocaust and World War II* (Lanham: Taylor Trade, 2002); Claudia Koonz, *The Nazi Conscience* (Cambridge, MA: Belknap Press of Harvard University Press, 2003); Richard Rhodes, *Masters of Death: The SS-Einsatzgruppen and the Invention of the Holocaust* (New York: Vintage Books, 2003); Heather Pringle, *The Master Plan: Himmler's Scholars and the Holocaust* (New York: Hyperion, 2006); and Doris Bergen, *The Holocaust: A Concise History* (Lanham: Rowman & Littlefield, 2009). The estimated number of victims killed is from McKale, *Hitler's Shadow War*, 454.

5. See *The Holocaust Chronicle: A History in Words and Pictures* (Lincolnwood, IL: Legacy Publishing, 2009), 699.

6. Koonz, *Nazi Conscience*, 1.

7. Peter Fritzsche, *Life and Death in the Third Reich* (Cambridge, MA: Belknap Press of Harvard University Press, 2008), 15.

8. Fritzsche, *Life and Death*, 5, 15.

9. David B. Clarke, Marcus A. Doel, and Francis X. McDonough, "Holocaust Topologies: Singularity, Politics, Space," *Political Geography* 15(1996): 457–89; at 458.

10. Roberto Esposito, *Bíos: Biopolitics and Philosophy*, translated by Timothy Campbell (Minneapolis: University of Minnesota Press, 2008), 4.

11. Koonz, *Nazi Conscience*, 3.

12. David Engel, *The Holocaust: The Third Reich and the Jews* (Harlow, UK: Longman, 2000), 24.

13. Burleigh and Wippermann, *The Racial State*, 23.

14. Clarke et al, "Holocaust Topologies," 459.

15. Michel Foucault, *"Society Must Be Defended": Lectures at the Collège de France, 1975–1976*, translated by David Macey (New York: Picador, 2003), 254.

16. Clarke et al., "Holocaust Topologies," 459.

17. Foucault, *"Society Must Be Defended,"* 254–55.

18. Foucault, *"Society Must Be Defended,"* 255.

19. J. S. Haller, Jr., *Outcasts from Evolution: Scientific Attitudes of Racial Inferiority, 1859–1900* (Carbondale: Southern Illinois University Press, 1971); Robert Miles, *Racism* (New York: Routledge, 1989); David N. Livingstone, *The Geographical Tradition: Episodes in the History of a Contested Enterprise* (Cambridge, MA: Basil Blackwell, 1992).

20. Sarah K. Danielsson, "Creating Genocidal Space: Geographers and the Discourse of Annihilation, 1880–1933," *Space and Polity* 13(2009): 55–68; at 59.

21. William H. Tucker, *The Science and Politics of Racial Research* (Urbana: University of Illinois Press, 1994), 26.

22. Saul Dubow, *Scientific Racism in Modern South Africa* (Cambridge: Cambridge University Press, 1995), 101.

23. Nancy L. Stepan, *"The Hour of Eugenics": Race, Gender, and Nation in Latin America* (Ithaca, NY: Cornell University Press, 1991), 25–27; Dubow, *Scientific Racism*, 121–22.

24. See for example Daniel J. Kevles, *In the Name of Eugenics: Genetics and the Uses of Human Heredity* (Berkeley: University of California Press, 1985); Stepan, *"The Hour of Eugenics"*; Stefan Kühl, *The Nazi Connection: Eugenics, American Racism, and German National Socialism* (Oxford: Oxford University Press, 1994); Edward J. Larson, *Sex, Race, and Science: Eugenics in the Deep South* (Baltimore: Johns Hopkins University Press, 1995); Diane B. Paul, *Controlling Human Heredity: 1865 to the Present* (New York: Macmillan, 1995); Nancy Ordover, *American Eugenics: Race, Queer Anatomy, and the Science of Nationalism* (Minneapolis: University of Minnesota Press, 2003).

25. Kühl, *Nazi Connection*, 84.

26. John R. Commons, *Races and Immigrants in America* (New York: Macmillan, 1907), 7.

27. L. H. M. Baker, *Race Improvement or Eugenics* (New York, 1912), n.p.

28. Stepan, *"The Hour of Eugenics,"* 28.

29. Michael Burleigh, *Death and Deliverance: "Euthanasia" in Germany 1900–1945* (Cambridge: Cambridge University Press, 1994), 3.

30. Paul B. Popenoe and Roswell Hill Johnson, *Applied Eugenics*, 2nd ed. (New York: Macmillan, 1918), 260.

31. Quoted in Tucker, *Science and Politics*, 60.

32. Kühl, *Nazi Connection*, 25.

33. Danielsson, "Creating Genocidal Space," 59.

34. Murray Edelman, *Political Language: Words That Succeed and Policies That Fail* (New York: Academic Press, 1977), 23.

35. Gearóid Ó Tuathail, *Critical Geopolitics* (Minneapolis: University of Minnesota Press, 1996).

36. Ordover, *American Eugenics*, 7.

37. Martin Glassner, *Political Geograpy*, 2nd ed. (New York: John Wiley, 1996).

38. Quoted in Mark Bassin, "Imperialism and the Nation State in Friedrich Ratzel's Political Geography," *Progress in Human Geography* 11(1987): 473–95; at 477.

39. Quoted in Danielsson, "Creating Genocidal Space," 63.

40. Esposito, *Bíos*, 16.

41. Esposito, *Bíos*, 16.

42. Danielsson, "Creating Genocidal Space," 66.

43. Danielsson, "Creating Genocidal Space," 62–63.

44. Danielsson, "Creating Genocidal Space," 55, 64–66.

45. Bassin, "Imperialism and the Nation State," 476.

46. Charles B. Davenport, *Eugenics: The Science of Human Improvement by Better Breeding* (New York: Henry Holt, 1910), 16.

47. Christopher R. Browning, *Nazi Policy, Jewish Workers, German Killers* (Cambridge: Cambridge University Press, 2000), 11.

48. Browning, *Nazi Policy*, 5.

49. Bergen, *The Holocaust*, 1.

50. Nikolas Rose, *The Politics of Life Itself: Biomedicine, Power, and Subjectivity in the Twenty-First Century* (Princeton, NJ: Princeton University Press, 2007), 24.

51. Jeff McMahan, *The Ethics of Killing: Problems at the Margins of Life* (New York: Oxford University Press, 2002), vii.

52. Giorgio Agamben, *Remnants of Auschwitz: The Witness and the Archive* (New York: Zone Books, 2002), 82.

53. Bergen, *The Holocaust*, 55.

54. Orlando Patterson, *Slavery and Social Death: A Comparative Study* (Cambridge, MA: Harvard University Press, 1982), 38–39.

55. Patterson, *Slavery and Social Death*, 41.

56. Clarke et al., "Holocaust Topologies," 459.

57. Esposito, *Bíos*, 10.

58. Bergen, *The Holocaust*, 52–53.

59. Burleigh and Wippermann, *The Racial State*, 44.

60. McKale, *Hitler's Shadow War*, 26.

61. Koonz, *Nazi Conscience*, 3.

62. Iris M. Young, *Justice and the Politics of Difference* (Princeton, NJ: Princeton University Press, 1990), 234.

63. Linda McDowell, *Gender, Identity & Place: Understanding Feminist Geographies* (Minneapolis: University of Minnesota Press, 1999), 100.

64. Stuart C. Aitken, *Family Fantasies and Community Space* (New Brunswick, NJ: Rutgers University Press, 1998), 133–34.

65. David Sibley, *Geographies of Exclusion: Society and Difference in the West* (London: Routledge, 1995), 3.

66. Amartya Sen, *Identity and Violence: The Illusion of Destiny* (New York: W.W. Norton, 2006), 2.

67. Koonz, *Nazi Conscience*, 5; see also Pringle, *The Master Plan*.

68. Bergen, *The Holocaust*, 62.

69. Proctor, *Racial Hygiene*, 131; Bergen, *The Holocaust*, 72.

70. Burleigh and Wippermann, *The Racial State*, 48–49.

71. Proctor, *Racial Hygiene*, 132.

72. Koonz, *Nazi Conscience*, 6–9.

73. Koonz, *Nazi Conscience*, 6.

74. Koonz, *Nazi Conscience*, 6–7.

75. Koonz, *Nazi Conscience*, 7.

76. Koonz, *Nazi Conscience*, 8.

77. Esposito, *Bíos*, 115.

78. Burleigh and Wippermann, *The Racial State*, 51.

79. Burleigh and Wippermann, *The Racial State*, 56.

80. Daniel J. Kevles, *In the Name of Eugenics: Genetics and the Uses of Human Heredity* (Cambridge, MA: Harvard University Press, 1995), 47.

81. Ordover, *American Eugenics*, 133.

82. Stepan, *"The Hour of Eugenics,"* 31; Paul, *Controlling Human Heredity*, 82–83.

83. Paul, *Controlling Human Heredity*, 84–91.

84. Proctor, *Racial Hygiene*, 101.

85. Quoted in Kevles, *In the Name of Eugenics*, 111.

86. Quoted in Paul, *Controlling Human Heredity*, 83.

87. Esposito, *Bíos*, 133.

88. Lifton, *The Nazi Doctors*, 46.

89. Proctor, *Racial Hygiene*, 178.

90. Quoted in Lifton, *The Nazi Doctors*, 47.

91. Burleigh, *Death and Deliverance*, 19.

92. Quoted in Richard A. Koenigsberg, *Nations Have the Right to Kill: Hitler, the Holocaust and War* (New York: Library of Social Sciences, 2009), 10.

93. Koenigsberg, *Nations*, 10.

94. Esposito, *Bíos*, 133.

95. Esposito, *Bíos*, 134.

96. Esposito, *Bíos*, 137.

97. Lifton, *The Nazi Doctors*, 63.

98. Friedlander, *Origins of Nazi Genocide*, 66–67.

99. Friedlander, *Origins of Nazi Genocide*, 67.

100. Friedlander, *Origins of Nazi Genocide*, 67.

101. Lifton, *The Nazi Doctors*, 25.

102. Burleigh and Wippermann, *The Racial State*, 136.

103. Proctor, *Racial Hygiene*, 102; see also Burleigh and Wippermann, *The Racial State*, 136–42; and Lifton, *The Nazi Doctors*, 22–44.

104. Lifton, *The Nazi Doctors*, 29.

105. Proctor, *Racial Hygiene*, 104.

106. Proctor, *Racial Hygiene*, 108–9; see also Pringle, *The Master Plan*, 264.

107. Proctor, *Racial Hygiene*, 108; Lifton, *The Nazi Doctors*, 27; Burleigh and Wippermann, *The Racial State*, 138.

108. Proctor, *Racial Hygiene*, 108–9.

109. Pieter Admiraal, "Euthanasia and Assisted Suicide," in *Birth to Death: Science and Bioethics*, edited by David C. Thomasma and Thomasine Kushner (Cambridge: Cambridge University Press, 1996), 207–17; at 207.

110. Admiraal, "Euthanasia," 208–9.

111. Admiraal, "Euthanasia," 210.

112. Jeff McMahan, *The Ethics of Killing: Problems at the Margins of Life* (Oxford: Oxford University Press, 2002), 457.

113. Burleigh, *Death and Deliverance*, 13.

114. See Koenigsberg, *Nations*, 11.

115. Lifton, *The Nazi Doctors*, 51.

116. Quoted in Lifton, *The Nazi Doctors*, 51; see also Proctor, *Racial Hygiene*, 185–86.

117. Proctor, *Racial Hygiene*, 186.

118. Proctor, *Racial Hygiene*, 186; see also Lifton, *The Nazi Doctors*, 52.

119. Proctor, *Racial Hygiene*, 187; Lifton, *The Nazi Doctors*, 54; Burleigh, *Death and Deliverance*, 100–101.

120. Proctor, *Racial Hygiene*, 188.

121. Proctor, *Racial Hygiene*, 187.

122. Quoted in Burleigh, *Death and Deliverance*, 127.

123. Quoted in Burleigh, *Death and Deliverance*, 112.

124. Quoted in Lifton, *The Nazi Doctors*, 63.

125. Burleigh, *Death and Deliverance*, 112–13.

126. Burleigh, *Death and Deliverance*, 4.

127. Burleigh, *Death and Deliverance*, 4.

128. Lifton, *The Nazi Doctors*, 65.

129. Proctor, *Racial Hygiene*, 190–91.

130. Burleigh and Wippermann, *The Racial State*, 144.

131. Proctor, *Racial Hygiene*, 189.

132. Lifton, *The Nazi Doctors*, 71.

133. Proctor, *Racial Hygiene*, 190.

134. Burleigh, *Death and Deliverance*, 144–49.

135. Burleigh, *Death and Deliverance*, 149.

136. Proctor, *Racial Hygiene*, 193.

137. Burleigh, *Death and Deliverance*, 253.

138. Burleigh, *Death and Deliverance*, 160. It should be noted, also, that the killings continued after the official stoppage. Additional asylums were used for mass killings, although the methods were less industrial; rather, officials relied upon lethal injections, starvation, and other forms of inexpensive means. See Burleigh, *Death and Deliverance*, 238.

139. Burleigh, *Death and Deliverance*, 180.

140. Bergen, *The Holocaust*, 167–68.

141. Bergen, *The Holocaust*, 168.

142. Laurence Rees, *Auschwitz: A New History* (New York: Public Affairs, 2005), 199.

143. Browning, *Nazi Policy*, 39.

144. Browning, *Nazi Policy*, 33.

145. Arad, *Belzec, Sobibor, Treblinka*, 2.

146. Rhodes, *Masters of Death*, 4.

147. Bergen, *The Holocaust*, 154.

148. Bergen, *The Holocaust*, 155. Nazi officials and officers of the *Einsatzgruppen* also experimented with other, more efficient and less traumatic (from the killers' perspective) forms of killing. Some units, for example, converted buses and moving vans into mobile gas chambers.

149. Rees, *Auschwitz*, 197.

150. It also bears mentioning that the mass shootings could bring too much attention to the atrocities being committed. While, on the one hand, Nazi officials welcomed the terror that resulted from mass spectacles of murder, on the other hand they were aware that such crimes might turn local populations against the Germans.

151. Friedlander, *Origins of Nazi Genocide*, 286.

152. Arad, *Belzec, Sobibor, Treblinka*, 11.

153. Friedlander, *Origins of Nazi Genocide*, 287.

154. Nikolaus Wachsmann, "The Dynamics of Destruction: the Development of the Concentration Camps, 1933–1945," in *Concentration Camps in Nazi Germany: The New Histories*, edited by Jane Caplan and Nikolaus Wachsmann (London: Routledge, 2010), 21.

155. Arad, *Belzec, Sobibor, Treblinka*, 1.

156. Wachsmann, "Dynamics of Destruction," 25–26.

157. Wachsmann, "Dynamics of Destruction," 24–25.

158. Bergen, *The Holocaust*, 114–15.

159. Bergen, *The Holocaust*, 112–13.

160. Rees, *Auschwitz*, 203.

161. McKale, *Hitler's Shadow War*, 255.

162. McKale, *Hitler's Shadow War*, 259.

163. Esposito, *Bíos*, 9.

Chapter Three

Starving for the State: China

Between 1958 and 1961, approximately 40 million people were "allowed" to die in the People's Republic of China's Great Leap Forward.[1] For the vast majority, their deaths occurred as a result of starvation and other opportunistic diseases: diarrhea, dysentery, fever, typhus. In villages and cities, farms and factories, China's citizens perished as a result of misguided policies and a political system cowed into submission. Those who dared to mention that millions of China's subjects were dying were themselves subject to public humiliation, beatings, imprisonment, and execution. Indeed, perhaps 2 million people died not from starvation, but from direct and brutal violence—sanctioned by the state.

In March 1959, as reports circulated that massive numbers of people were dying, Mao Zedong made a calculated valuation of life. He explained that "when there is not enough to eat people starve to death. It is better to let half of the people die so that the other half can eat their fill."[2] Many readers have no doubt experienced pangs of hunger. Few have endured prolonged starvation. What does it mean to starve to death?

In dry medical terms, starvation is "a state in which the energy demands of an organism exceed supply, forcing reliance on endogenous reserves."[3] In more plain language, when the human body does not receive an adequate intake of nutrients, it begins to utilize glycogen, fats and proteins stored in the body. And in fact this process—called *incipient starvation*—begins four to six hours after one's last meal. First, liver glycogen is broken down into glucose to nourish the body. However, there is only enough glycogen stored in the liver to last a few hours; if no external nutrients are provided, the body next turns to fat reserves as a source of energy. In general, life can be sustained for one to three months—a period known as *acute starvation*. When fatty reserves are depleted, the body next begins to deplete proteins,

81

particularly those in muscles. At this point, cell functions degenerate and organs begin to atrophy. When weight falls to about two-thirds normal, death is considered inevitable.[4]

As one starves to death, the body progressively weakens and gradually wastes away. People become lethargic, listless, and withdrawn. Fluids, leaking internally from blood vessels and from decomposing tissues, begin to collect under the skin. This causes the distinctive swelling of the face, feet, legs, and abdominal region and provides a startling contrast to the otherwise emaciated body: ribs poking through chests, arms and legs skinny as twigs, vacant, bulging eyes sitting above sunken cheeks. In such weakened conditions, many people die from infectious diseases before they actually starve to death.

The famine in China (1958–1961) is often not considered genocidal. Many scholarly accounts of genocide do not even include Maoist China.[5] Why not? Why the omission? It certainly isn't the death toll. During these four years, approximately 40 million Chinese citizens died because of identifiable policies and practices initiated by the state. Furthermore, as recent scholarly accounts demonstrate, Chinese authorities were well aware of the massive number of deaths.[6] Valentino acknowledges that "many scholars have concluded that these deaths were almost entirely the *unintentional* consequences of ill-conceived or poorly implemented social and economic programs, rather than systematic mass killings."[7] However, our interpretation of mass death in China hinges exactly on the biopolitical and ethical distinction between "killing" and "letting die"—a distinction that is not adequately considered vis-à-vis the study of genocide. Indeed, most discussions follow the *politically derived* definition of genocide as forwarded by the United Nations Genocide Convention. As stipulated in Article II, genocide is defined as "any of the following acts committed with *intent* to destroy, in whole or in part, a national, ethnical, racial or religious group, as such: (a) Killing members of the group; (b) Causing serious bodily or mental harm to members of the group; (c) Deliberately inflicting on the group conditions of life calculated to bring about its physical destruction in whole or in part; (d) Imposing measures intended to prevent births within the group; (e) Forcibly transferring children of the group to another group."[8]

The 40 million deaths that occurred in China between 1958 and 1962 are thus problematic on two counts. On the one hand, it is not clear that any particular "group" was targeted. To this, some have argued that certain ethnic minorities—notably the Tibetans—were set apart; other nonethnic groups, such as "counterrevolutionaries," were likewise targeted. On the other hand, and following Valentino, is the assertion that the deaths were not intentional. Recently, for example, Yang and Su have argued that "Mao Zedong and his

colleagues at the pinnacle of the Chinese party-state should take major responsibility for the Great Leap Famine. Nevertheless, it is also true that Mao
Zedong, Zhou Enlai, and other Chinese leaders did not *intend* to kill millions
of peasants by mass starvation" (emphasis added).[9]

In the case of Nazi Germany, "intent" is blatantly obvious, so much so that
the Holocaust remains the prototypical "case" of genocide. Consequently,
cases such as the massive loss of life that occurred in China remain less clear
and subject to debate. In this chapter I argue that the famine cannot—should
not—be simply swept aside as an unfortunate consequence of ill-conceived
or poorly executed policies. Consider that in 1957 Mao announced: "Let us
imagine how many people would die if war breaks out. There are 2.7 billion
people in the world, and a third could be lost. If it is a little higher it could be
half. . . . I say that if the worst came to the worst and one-half dies, there will
still be one-half left, but imperialism would be erased and the whole world
would become socialist. After a few years there would be 2.7 billion people
again."[10] More than just a geographical fantasy to construct a global socialism, what is most salient is Mao's seeming disregard for *all* human life. As I
conclude in this chapter, to "let die" is in fact on par with the more deliberate
and direct act of killing.

POWER/KNOWLEDGE/VIOLENCE IN MAOIST CHINA

In the many writings of Michel Foucault, knowledge assumes a central place.
Indeed, a dominant motif of Foucault's work was to provide a critique of the
way modern societies control and discipline their populations by sanctioning
the knowledge claims and practices of the human sciences.[11] In other words, a
Foucauldian approach would be concerned with how knowledge is produced
and subsequently deployed: knowledge normally derived from efforts of
academics, policy makers, and so forth; knowledge produced via research,
personal observation, and various forms of "data collection." Maoist China
reflects something different—a difference, I maintain, that is crucial for our
understanding of the Great Famine as genocide.

For Foucault, knowledge is inseparable from power. Foucault explains that
power and knowledge directly imply one another; there is no power relation
without the correlative constitution of a field of knowledge, nor any knowledge that does not presuppose and constitute at the same time power relations.
Power as conceived by Foucault, moreover, differs from traditional accounts.
In *Discipline and Punish*, Foucault asserts that power is exercised rather than
possessed.[12] Later, he elaborates that "power must be analyzed as something
which circulates, or rather as something which only functions in the form of

a chain. It is never localized here or there, never in anybody's hands, never appropriated as a commodity or piece of wealth."[13]

The forwarding of power as exercised is significant in that it directs our attention away from a singular dominant class or institution—or even the "state." Power and the production of knowledge are not simply the result of an oppressive system or set of apparatuses. On this point, Foucault is clear: power is not the privileged domain of a dominant class; authorities do not have a monopoly on the exercise of power, or on the production of knowledge.[14] Foucault explains that power is not something that is acquired, seized, or shared; relations of power are not in a position of exteriority with respect to other types of relationships (e.g., economic processes, knowledge relationships); there is no binary and all-encompassing opposition between rulers and ruled at the root of power relations; there is no power that is exercised without a series of aims and objectives; and where there is power, there is resistance, and this resistance is never in a position of exteriority in relation to power.[15]

Here it is useful to clarify the distinction between power and violence. As noted earlier, power—from a Foucauldian perspective—is relational; power is exercised only over free subjects, with "free" meaning individual or collective subjects who are faced with a field of possibilities in which several forms of conduct, several ways of reacting, and several modes of behavior are possible. For Foucault, power thus exists on two conditions, the first being that "the other" is recognized and maintained by a subject who acts, and the second being that, faced with a relationship of power, a whole field of responses, reactions, results, and possible interventions may emerge. Power is thus not simply a matter of consent, but it is also not a renunciation of freedom—a transfer of rights from one to another.

An exercise of violence, conversely, is totalizing. When violence is applied to a body, subjugation is complete. It removes the possibilities for active subjects to reinscribe themselves; it removes the possibility for resistance.[16] To this end, Barker explains that "violence involves a direct application of force upon the body of the other, reducing every possibility for independent action. Violence is applied directly to a body, but more than this it is applied to a body which is not recognized as being in a 'relationship' that would allow it to act autonomously."[17]

What are we to make of Maoist China? And what does Maoist China say about the power/knowledge coupling that literally *informed* the massive loss of life that occurred between 1958 and 1962? I suggest, in an extension of Foucault, that the power/knowledge relationship must be reconfigured as one of violence/knowledge. That in Maoist China, the circulation of knowledge was brought to a standstill, enforced by the state's monopoly on violence.

On October 1, 1949, the People's Republic of China emerged as a sovereign state.[18] More significant, however, is Judith Shapiro's observation that from 1949 to 1976—the year of Mao's death—Mao and the Communist Party sought to reengineer Chinese society by remolding human nature.[19] In short, Mao was instrumental in bringing to life—through death—a particular geographic imagination of what a communal utopia should be. However, unlike Adolf Hitler, who required *lebensraum* for the Nazi state and German *Volk*, Mao worked from within.

And in a further contrast to Hitler, the role of intellectuals in Maoist China was silenced. Thus, whereas the Nazi state enlisted the help of trained professionals—doctors, lawyers, engineers, among others—to carry out its policies, Mao took the opposite path. In fact, not only did Mao distrust intellectuals, he also held contempt for "local knowledges"—the knowledges gained from experience by peasants working the land for generations. As Shapiro writes, "numerous campaigns suppressed [both] elite scientific knowledge and traditional grass-roots practices, . . . stifling dissent through political labels, ostracism, and labor camp sentences." Trained scientists, who might utter words of dissent or caution, were often exiled or persecuted to death. Peasants, likewise, were treated harshly if they deviated from party lines.[20] Mao made clear his attitude toward intellectuals when he quipped that "bourgeois professors' knowledge should be treated as dogs' fart, worth nothing, deserving only disdain, scorn, contempt."[21]

Mao's violence/knowledge coupling is vividly illustrated in the Hundred Flowers movement. In the fall of 1955 Mao apparently called for a relaxation in the dogmatism and formalism that characterized the intellectual environment of China. With the slogan "Let a hundred flowers bloom, let a hundred schools of thought contend" Mao initiated a period in which free expression was allowed to flourish—that scholars could enter into academic debate without fear of persecution. Such debate extended to that of politics—and was formally codified in a speech delivered by Mao in February 1957.[22] In part, this relaxed environment appeared to be a consequence of an earlier effort of Mao to push through rapid economic changes—an event subsequently labeled the "Little Leap Forward." Those efforts failed, and many party officials had urged caution in introducing new policies.

By June 1957 the bloom had fallen off the flower. An anti-rightist purge was launched, which, over the next few years, silenced tens of thousands of scientists, engineers, journalists, intellectuals, and others.[23] Many scholars, then and now, maintain that the Hundred Flowers movement was a trap, a diabolical scheme designed to smoke out disloyal members of Chinese society.[24] According to this line of reasoning, Mao was convinced that China's inability to progress quickly—the "Little Leap"—was a failure not of policy but of people. He railed against what he perceived as widespread intransigence and outright resistance. Mao, in response, launched the Hundred Flowers campaign to identify those who disapproved of his policies—both intellectuals and other party members who had complained about Mao's agricultural policies.[25]

Whether the campaign was a trap or not is, at a certain level, irrelevant. Judith Shapiro is correct in her assessment that the "popular perception that it was a trap may be more significant than the reality." The salience is that "potentially unruly thinkers were *disciplined* and brought into line."[26] The ever-presence of violence altered the relationship between power and knowledge. Increasingly, throughout Mao's rule, power was subsumed by violence. Knowledge no longer circulated, as with power, but instead was locked into Mao's state of violence.

The first signs of famine began to appear in early 1958; by 1959 it was obvious—to those who chose to look—that famine was widespread. Preliminary reports, cautiously prepared, suggested that upwards of 5 million people were suffering from edema and that 70,000 people had already died. Medical doctors, however, were not permitted to diagnose edema or other famine-related symptoms. Instead, edema was euphemistically called "swollen sickness" (*fuzhong bing*) or "water illness" (*shuizhong bing*). In some places, notably prison camps, edema was simply referred to as Number Two Disease.[27] Consequently, while Mao and other party officials received reports that detailed horrendous conditions in the countryside, that millions of peasants were starving to death, that peasants were even resorting to cannibalism, no action would be taken. And later, when some officials would, however cautiously and guardedly, speak out, they would be publically humiliated—if not executed. Significantly, even executions in Maoist China reflected a deliberate calculation. As Dikötter explains, death "by execution, like everything else in the planned economy, was a figure, a target to be fulfilled, a table of statistics in which the numbers had to add up." In 1960, for example, Xie Fuzhu, minister of public security, announced that 4,000 people should be killed.[28]

In 1959, as the famine was deepening, two men in particular dared to speak out. One was Peng Dehuai, China's minister of defense. Significantly, Peng himself had suffered through starvation and knew firsthand the sufferings that famine entailed. Peng blamed Mao, and he would later pay for his comments. Also outspoken was Zhang Wentian, vice-minister of foreign affairs. Zhang argued that production targets were too high; that the claims of record crop yields were false; and that people were starving. And he rebuked those in authority who failed to take notice.[29]

Those who spoke up, like Peng and Zhang, were accused of harboring rightist sentiments; they were subjected to public humiliations, beatings, imprisonment, and execution. More broadly, fearing a widespread conspiracy—and using terror and violence to maintain state control—a series of purges was initiated by Mao. In Beijing alone, thousands of top officials were targeted by the end of 1959, including 300 high-ranking officials. In the provinces, those authorities who had attempted to minimize the effects of the famine were removed from their positions. By March 1960, some 190,000 people had been denounced and humiliated in public meetings; approximately 40,000 cadres were expelled from the party. This latter figure includes over 150 top-ranking provincial officials.[30] Overall, approximately 3.6 million party members were labeled or purged as rightists—while overall party membership *increased* from just under 14 million to over 17 million in the span of a year. As Dikötter concludes, whatever remnants of reason had managed to survive the folly of the Great Leap Forward were swept aside in a frenzied witch hunt that left farmers more vulnerable than ever to the naked power of the party. At every level, from the province, to the county, commune, and brigade, brutal and bloody purges were carried out, replacing lackluster cadres with hard, unscrupulous officials who trimmed their sails to benefit from the radical winds blowing from Beijing.[31]

THE POLITICAL MAKING OF FAMINE

The famine that gripped China between 1958 and 1962 resulted in the single largest loss of life in the twentieth century; upwards of 40 million people died. This translates to a loss of life of over 27,000 people *per day*, over 1,100 people *per hour*. And these deaths did not result from the carnage of war, of two sovereign states clashing over borders or resources. Rather, these deaths were self-imposed, in that it was the Chinese state that created the conditions leading to mass death.

The cause or causes of the famine have long been debated. Initially, of course, as the famine was under way, any reports of mass starvation were silenced. Gradually, as the conditions could no longer be ignored, party officials blamed nature—droughts, floods, and so forth. The peasants themselves were increasingly blamed—a topic I address toward the end of this chapter. Most accounts, however, forward a combination of structural factors, including agricultural policies, the promotion of industry, and international trade patterns. Curiously, many of these discussions fail to adequately consider the role of both discipline and violence in accounting for the massive death toll.

In this section I introduce a series of structural conditions that contributed to the famine; these are largely agreed upon in most accounts. First, however, I situate these conditions within the broader geographical imagination expounded by Mao, an imagination that dominated Chinese policy decisions at the time. To this end, Judith Shapiro provides a useful thematic framework to understand Maoist China. She explains that four core themes effectively capture the "historical factors, ideological influences, utopian dreams, and coercive political structures" that characterize the period under study.[32] The first theme is that of *political repression*. As the preceding discussion of the violence/knowledge coupling makes clear, the repression of intellectuals, scientists, officials, and ordinary people was pervasive.[33] A second theme is that of *utopian urgency*. In effect, a particular geographical imagination of a modern, communist utopia was envisioned by Mao—a vision adopted by local leaders and party officials. However, this utopia was to be achieved immediately. It was the philosopher Hegel who remarked that progress, like evolution, comes in sudden leaps and bounds. It was from this sentiment that Mao would call his program "The Great Leap Forward."[34] Third is a theme of *dogmatic uniformity*: the imposition of "one-knife-cuts-all" sentiments that ignored regional geographic variations and local practices.[35] Last is a theme of *state-ordered relocations*, or reconfigurations of society by administrative fiat.[36]

Combined these four themes coalesce around Mao's quest to achieve a communist utopia, with Mao himself as the supreme leader. And similar to Hitler's megalomaniacal dream of world-building—a world free of Jews and other "undesirable" populations, a world dominated by racially pure Aryans—Mao held a similar though less racially informed vision. For Mao, like Hitler, populations were nothing more than means to an end—bodies to be sacrificed or annihilated for the state. Mao's philosophy held that through concerted exertion of human will and energy, material conditions could be altered and all difficulties overcome in the struggle to achieve a socialist utopia.[37]

Mao's *will to space* was predicated on the discipline of people and the regulation of populations to work harder, to work longer, to sacrifice more, and to expect less. And in the end, China would not only become a land of abundance, China would also surpass all other countries—both communist and capitalist—as the preeminent world power. It was thus in October 1957, with the "support" of his party, that Mao uttered the slogan that crystallized his vision: "Greater, faster, better and more economical."[38] Thus was born the Great Leap Forward.

Mao was joined, not surprisingly given the violence/knowledge coupling of China, by most top officials. Li Fuchun, head of the State Planning Commission, was one of the first economic planners to support Mao's utopian dream. Other early supporters were Liu Shaoqi, state president, and Zhou Enlai, premier. Other officials, such as Li Xiannian, minister of finance, and Bo Yibo, chair of the State Economic Commission, had opposed the Little Leap Forward but quickly sensed the direction of the prevailing winds and came on board.[39] Combined, these party officials, along with a legion of loyal party members at the local level, contributed to an overall calculated management of life. As Dikötter explains, "The whole country became a universe of norms, quotas and targets from which escape was all but impossible, as loudspeakers blasted slogans, cadres checked and appraised work, and committees endlessly ranked and rated the world around them. And classification of individual performance would increasingly determine the kind of treatment meted out—down to the ladle of gruel in the canteen in times of hunger."[40]

In Nazi Germany, Hitler demanded *lebensraum*—new territories to be conquered, occupied, and exploited for the requirements of his expanding state. In China, conversely, Mao looked inward. To this end, Mao wanted to raise the output of China's citizenry, but to do so with minimal capital investment.[41] Subsequently, Mao advocated labor-intensive methods, forcing millions of peasants into the fields and factories as one might deploy an army. And indeed it was an army: an army of surplus laborers to be exploited by the state in pursuit of a utopian dream.

Agricultural productivity was the key to China's rapid development—a conclusion of Mao that was heavily influenced by Soviet policy. Given that approximately 80 percent of China's population worked in agriculture, party officials throughout the 1950s viewed grain procurement as the key for overall development. Quite simply, farmers would plant, harvest, and sell their grain to the state through a practice of procurement; the state, in return, could either sell back surplus grain to the peasants, or exchange the grain on the international market for currency or commodities. In the years *prior to* and *during* the Great Leap, most of the grain actually produced in China was used to finance industrialization or any other program determined by Mao and the party.[42] Throughout the 1950s and 1960s, over half of China's exports to the Soviet Union consisted of agricultural commodities: fibers, tobacco, grain, soybeans, fruit, oils, rice. China was also "dumping" products throughout Africa and Asia—bicycles, sewing machines, fountain pens; these commodities were sold below cost simply to "show" that China was ahead of the Soviet Union.[43] China also used its agricultural "surpluses" to pay off outstanding loans—most notably, again, to the Soviet Union. Moreover, throughout the years of the famine, Mao decided to initiate an accelerated repayment plan, although the Soviet Union never requested such a time schedule. This would form one of the most enduring myths associated with the famine—that China's hunger was a direct result of Moscow's heartless pressure to immediately pay back debts, thereby "forcing" China to increase its grain procurements.[44]

In actuality, the practice of state grain procurements began soon after the founding of the People's Republic of China. In 1953 central planners introduced a monopoly on grain. All farmers were required to sell any surplus yields to the state at prices determined by the state. The aim, ostensibly, was to stabilize the price of grain. In actuality, this was a means for the state to amass profit. A basic grain ration was established, whereby farmers were allotted a monthly ration of 13 to 15 kilos of seed, fodder, and basic grain per head. However, 23 to 26 kilos of unhusked grain were required to provide the approximately 1,700 to 1,900 calories per day considered to be the bare minimum for survival. As Dikötter concludes, the notion of surplus was a political construct designed to give legitimacy to the extraction of grain from the countryside; villagers were forced to sell grain before their own subsistence needs were met. Any extra grain needed—for basic survival—had to be bought back from the state by villagers. Payment though was in the form of "work points," as wages were gradually abolished.[45] In short, the practice of procurement, with the inhumanly low ratios established, almost ensured that some level of malnutrition or starvation would occur.

Party officials also understood that the system required ever-increasing yields. However, these were to be achieved not through the input of capital, but rather through energetic labor and innovative, nonbourgeois methods. Notable among these were the practices of "close cropping" and "deep plowing." Here, Mao drew inspiration not from Chinese scientists and intellectuals, but from Soviet scientists, such as Trofim Denisovich Lysenko, an individual who completely dominated agricultural science in the Soviet Union. So dominant was his authority, that those who opposed Lysenko were executed or imprisoned. Lysenko did not "believe" in genetics; he dismissed this so-called science as an "expression of the senile decay and degradation of bourgeois culture" and resolutely rejected the "fascist" theories that plants and animals have inherited characteristics that selective breeding could develop. Thus, unlike the German scientists who dominated the Nazi state, Lysenko followed Lamarck, asserting that environmental factors were all-powerful in the shaping of organic structures and characteristics. It was possible, according to Lysenko, that under the proper environmental conditions, orange trees could be transformed into apple trees.[46]

It was from Mao's application of Soviet agricultural science that many of the practices initiated throughout China were first proposed. It was Lysenko, for example, who argued that seeds and saplings should be planted in close proximity—close cropping—because, according to his "law of the life of species," individuals of the same species do not compete but help each other survive.[47] Such reasoning readily conformed to Mao's understanding of class struggle. Mao reasoned that "with company they grow easily, when they grow together they will be more comfortable."[48] Terenty Maltsev, another Soviet expert, maintained that yields could be increased and soil textures improved through a practice of "deep plowing." The deeper the planting, according to Maltsev, the stronger the roots and the taller the stalks.[49] Throughout China, peasants were forced to dig ever-deeper furrows—some as deep as ten feet.

Farmers, of course, knew better than to practice deep plowing or close cropping. Having tilled the land for generations, the women and men whose livelihood depended on the harvest attempted to reason with party officials. But their pleas came to nothing. As discussed later, it was not uncommon for farmers who objected to the state-mandated practices to be beaten or punished.[50] Indeed, as Becker concludes, with the full power of the party behind the orders, the peasants had no choice but to obey.[51] In short, traditional farming practices were done away with in favor of untested methods advocated by the state.[52]

Figure 3.1. Propaganda photograph of peasants harvesting rice in the province of Guangdong during the Great Leap Forward, c. 1958. *Source:* Getty Images

Figure 3.2. Propaganda photograph of women harvesting cocoons from silkworms during the Great Leap Forward. *Source:* Getty Images

When harvest time arrived, local officials declared that their regions had indeed produced record yields; Mao's propaganda machine then publicized the claims with great fanfare.[53] Photographs were produced that showed wheat fields so abundant that children could actually stand on the top of the stalks. These photographs, and many others, however, were manipulated, just like the false statistics that were reported up the chain of command.

As record yields were announced, the proportion of grain procured by the state likewise increased. As one local cadre explained, "In 1960 we were given a quota of 260 tons. This was increased by 5.5 tons a few days later. Then the commune held a meeting and added a further 25 tons. After two days, the commune phoned us to say that the quota had gone up to 315 tons."[54] Furthermore, the state was unwilling to sell back grain—even if farmers had acquired work points. Rather, tons of grain were exported overseas to either pay for new technologies or to repay debts.

The nefarious relationship between poor farming techniques, exaggerated claims, and increased state procurements contributed significantly to the famine. In the autumn of 1959, the grain harvest *dropped* by at least 30 million tons as compared to 1958; officials, however, reported much higher yields. State procurement targets were set at 40 percent—the highest level set. Normally, state procurements ranged between 20 and 25 percent.[55] Chang and Halliday conclude that for the Chinese population, "the Great Leap was an enormous jump—but in the amount of food extracted. This was calculated on the basis, not of what the peasants could afford, but of what was needed for Mao's program."[56]

To increase the efficiency of agricultural labor, Chinese officials set about to transform the entire social structure of the farmers' traditional way of life. The collectivization of agriculture was a central goal of the Chinese Communist Party long before it gained power in 1949. In the early years, as the party solidified its control over the rural populations, discipline was facilitated through violent practices of land reform. Although the stated purpose of land reform was to increase agricultural productivity and to redress rural inequalities in land ownership, the underlying purpose was simply to remove potential political opponents, namely, the landlords themselves. Thus, lands were taken, and owners were punished or killed. The definition of a landlord was highly malleable, depending on political necessity.[57]

Following the lead of the Soviet Union, Chinese officials—including Mao Zedong, Zhou Enlai, Li Dazhao, Chen Duxiu, and Liu Shaoqi—believed that the key to economic development and China's future industrialization was dependent upon a strictly disciplined spatial organization of society. Collectives were to fit this requirement. As a form of social organization, collectives were in principle to rationalize production, thereby creating surpluses that could be used to develop industry.[58]

According to Xin Meng and colleagues, land reform policies and the collectivization of social life occurred in three phases. The first, which began in 1952, required farmers to form mutual aid teams. Compared to later collectives, these were very small in scope, composed of between six to nine households. Within these "teams," households were expected to pool their assets and land. A second phase, beginning around 1954, has been termed "low collectivization." Now, all households in a village were required to combine their land and assets; in return, each household was entitled, in principle, to the amount it contributed to the overall collective. It was at this point that peasants were forced to sell a quota of their grain to the state at a set low price; any surplus beyond that procured by the state could be sold in markets.[59] In general, the state procured between 20 and 25 percent of the output.

A third phase, known as "high collectivization," began in the late 1950s. Under this practice, although farmers in each village had contributed land and capital assets for production, the amount they received in return depended only on their labor input. In effect, private property rights to land and assets were eliminated.[60] Collectivization was implemented with extreme haste in 1955 and 1956; by the middle of 1956, approximately 90 percent of Chinese peasants had been moved into collectives.[61]

This period of collectivization was epitomized by the establishment of *people's communes*—massive enclosures of people. The first commune established in April 1958, located at Chayashan, in Henan province, combined 27 collectives, 9,300 farms, and 43,000 people. As the party announced, "a new social organization appeared fresh as the morning sun above the broad horizon of East Asia."[62] Over the next few months, communes proliferated; by October 1958 most provinces had announced that they had completed the transition to people's communes.[63] Approximately 90 percent of all existing collective farms were now grouped together into massive communes—some with memberships exceeding 100,000 people.[64]

The "creation of the communes," as Valentino writes, "facilitated the imposition of radical communist social policies and the unprecedented regimentation of peasant life by the state."[65] Here we see the "prosaics of state-craft" so well enunciated by Joe Painter. Here we see the multitudinous means by which the Chinese state intervened into daily life—and death—of the peasants. Everything was to be communal: land, draft animals, tools. There was even talk of collectivizing wives.[66] In terms of infrastructure, people would no longer live in houses, but rather in huge dormitories. The elderly would live in "Happiness Homes," while children would be raised, communally, in state-run nurseries. Even eating was to be communal; private kitchens were banned, to be replaced by huge communal mess halls. Peasants were not only banned from eating at home, their woks and stoves were smashed.[67]

In practice, most of this infrastructure failed to materialize. Houses were often razed—both to salvage wood for fuel for the furnaces or to clear space for dormitories—but the dorms themselves were rarely constructed. As a result, many people found themselves crammed into small huts, or sharing tiny apartments with little or no running water or electricity. The communal kitchens, however, were established—but not to provide adequate nutrition.

By the fall of 1958, more than 2.65 million commune mess halls were rushed into operation, with an estimated 70–90 percent of the rural population eating in these.[68] As Chang and Halliday identify, "total control over food gave the state a terrifying weapon."[69]

Within the communes, wages likewise were abolished, replaced by a short-lived but elaborate credit system, devised, ostensibly, so that workers would receive "wages" according to their abilities. In effect, the system epitomized the calculations underlying the valuation of life within Maoist China. Becker explains that the work point system constituted a valuation of a person based on "ability" as opposed to "need"; it was a system designed, in principle, to achieve an egalitarian distribution of wealth—not equal, but egalitarian. Becker describes the cumbersome system:

> First, each peasant was graded on a scale of 1–10, according to the individual's physical strength and health. Then each type of work was graded separately with different accounting systems for each crop, for the various jobs in the construction of dams and irrigation works, and for sideline work such as raising chickens or repairing tools. . . . [Initially] a contract system was employed, under which each peasant, production team or brigade agreed to carry out a fixed amount of labor in a given period to meet the production targets set by the state. Surplus output was then divided among the different levels according to another complicated formula.[70]

Over time, the work point system was discontinued and peasants received next to nothing. Surpluses were rarely, if ever, given back to the commune. Rather, surplus crops were kept by the state to be distributed to the urban areas or, more commonly, exported overseas at grossly reduced rates—or free of cost.[71]

Most peasants, perhaps not surprisingly, opposed being forced to live in collectives—let alone communes. As a result, Chinese officials introduced widespread disciplinary practices. Discipline, according to Michel Foucault, is a form of power, "a modality for its exercise, comprising a whole set of instruments, techniques, procedures, levels of applications."[72] Disciplinary techniques, moreover, whether appropriated by prisons, schools, the military, or *collectives*, share a number of characteristics. Disciplinary techniques are dual-sided, in that they simultaneously enable and repress, organize and atomize.[73] In other words, these practices are initiated to make bodies both productive and docile. Indeed, discipline is always directed at the body. As Foucault writes, "it is always the body that is at issue—the body and its forces, their utility and docility, their distribution and their submission."[74] Foucault continues that "the body is . . . directly involved in a political field; power relations have an immediate hold upon it; they invest it, mark it, train it, torture it, force it to carry out tasks, to perform ceremonies, to emit signs."[75] Discipline produces subjected and practiced bodies.[76]

Within the communes, canteens, and work sites, discipline was pervasive, amounting to a form of social death. At one point, Mao even contemplated eliminating people's names, to be replaced with numbers. And in fact, in some areas peasants were sent to work with numbers sewn on the back of their clothes. According to Chang and Halliday, "Mao's aim was to dehumanize China's 550 million peasants and turn them into the human equivalent of draft animals."[77]

The subjugated body, the productive body, the docile body, the socially dead body, is in part the result of an exercise of power. Philip Barker, however, raises an interesting question that speaks to the practices enacted during the Great Leap Forward. Barker asks: "If we are all implicated in practices of domination and subjection, then how is it possible to distinguish between different applications of these practices?" He continues, "Does this not leave us unable to distinguish between the dominating practices that a parent may apply to their child, the dominating practices of the riot police against dem-

Figure 3.3. Propaganda photograph of women operating an irrigation system during the Great Leap Forward, c. 1958. *Source:* Getty Images

onstrators, and the dominating practices of torture/death camps?" Barker, consequently, questions how it is possible to distinguish between "practices of domination and subjection" and an "outright engagement in unmitigated violence."[78]

In Maoist China, this transformation materialized in the body of the starving peasant. When discipline proved insufficient—because the people were literally being worked and starved to death—other forms of coercion were required. As peasants weakened from exhaustion and starvation, local cadre beat and punished the peasants, accusing the dying workers of abrogating their responsibility, of being disloyal to the party, the state, and most especially, to Mao. On August 19, 1958, Mao instructed his provincial chiefs: "When you order things handed over [i.e., grain procurements] and they are not handed over, back up your orders with force."[79] As Becker explains, "Party cadres had increasingly to rely on force and terror to get the peasants to obey their orders. At the height of the famine, they wielded the power of life and death because they controlled the grain stores and could kill anyone by depriving them of food."[80]

Punishments—including imprisonment and torture—were both brutal and bizarre. Those accused of harboring rightist sentiments, as indicated above, were arrested and sent to prisons and labor camps. So too were those accused of failing to perform adequately at their tasks. Other tortures and punishments were utilized for those accused of theft. Men, women, and children were publically humiliated—forced to wear dunce caps and paraded through the streets; cardboard placards were hung around people's necks, indicating their "crimes" for all to see. Hair was torn out; ears and noses were cut off; testicles beaten; people were branded; husbands and wives forced to beat each other.[81] Some people were covered in feces and urine; others were forced to eat these excrements. Torture, as Paul Kahn explains, is a mark of sovereignty.[82] As such, torture should be conceived as a technique of state power.[83] Foucault maintains that torture forms part of a ritual, one that includes two components. First, it must mark the victim; it is intended to physically or symbolically brand the victim with guilt. Thus the humiliation: the covering of bodies with feces and urine. The sign is clear: *you are shit*. Second, the public torture (and possible execution) must be spectacular. It must be seen by all as a triumph of the state.[84] Not to be hidden, but to make an individual, singled out from the masses, as a lesson.

Other punishments were to kill—the ultimate authority of a sovereign power. However, these likewise were often spectacles, to be used as means of enforcement. People were buried alive, or locked in cellars and left to die.[85] Others were "simply" shot. No matter the practice, the message remained constant. There is no life beyond that determined by the state.

As we have seen, the attempt to rapidly increase agricultural productivity was *not* to feed the country's growing population. Rather, agricultural "surpluses" were the necessary ingredient for Mao's dream of becoming *the* world leader in industry. In 1957 Mao boldly predicted that China would outpace the then world's leading steel producer—the United Kingdom—within fifteen years. Dikötter describes Mao's utopian fantasy of rapid industrialization: "Smoking factory stacks, whirring machine tools, the hooting of factory whistles, towering blast furnaces glowing a deep red with fire: these were the consecrated images of a socialist modernity."[86] Dikötter concludes that "Mao may not have been an expert on industry, but he seemed able to rattle off the steel output of virtually every country at the drop of a hat."[87]

Similar to the calculations of agriculture, production quotas were established and reestablished, as loyal party members bolstered their own good standing. Production quotas were ever-moving targets. In 1957 the target for steel production was 5.35 million tons; by May the number climbed to 8.5 million tons; and by June it skyrocketed to over 10 million tons. And within twenty years, Mao predicted that China would be producing well over 700 million tons.[88]

The key to China's industrialization was found in the "backyards" of peasant-farmers. By bringing industry to labor rather than the other way around, peasants were to farm during the day and to produce steel at night. At the height of the campaign, approximately 90 million peasants were forced to construct and work these furnaces.[89] Built of stone, sand, clay, or bricks, these furnaces were "some three or four meters high with a wooden platform on top, supported by beams. A sloping ramp provided access to the furnace, farmers scuttling up and down with sacks of coke, ore and flux on their backs or baskets slung on long poles. Air was blown through the bottom, the molten iron and slag being released through tap holes."[90]

To satisfy the demands of the furnaces, local authorities required that all households contribute their pots and pans, shovels and hoes, to be melted down to produce Mao's steel. Throughout the countryside "the population was coerced into donating virtually every piece of metal they had, regardless of whether this was being used in productive, even essential objects."[91] According to Dikötter, "the leadership got its record, although much of it was slag, unwashed ore or mere statistical invention; iron ingots from rural communes accumulated everywhere, too small and brittle to be used in modern rolling mills."[92] Mao himself acknowledged that only 40 percent of the steel produced was of acceptable quality; that more than 3 million tons were essentially useless.[93]

Backyard furnaces, moreover, were exceptionally wasteful, both in terms of finances and natural resources. On the one hand, one ton of iron from a backyard furnace was estimated to cost 300 to 500 *yuan*, twice the amount needed by a modern furnace, to which had to be added four tons of coal, three tons of iron ore, and thirty to fifty working days.[94] On the other hand, the search for fuel to feed the incessant furnaces led to massive devastation. Forests were cleared for fuel; hills were dynamited in search of coal. And when these supplies were exhausted, or inaccessible, entire villages were destroyed to be used as fuel. In one commune alone, over fifty thousand buildings were destroyed—homes razed to the ground—to be used to stoke the fires of the ineffective furnaces.[95]

Such callous disregard for nature, however, was in line with the other component of Mao's Great Leap Forward: to conquer nature and spread agriculture, again, to feed industry. Mao's vision was to literally turn sand into seed, as deserts would be transformed from barren wastelands into lush and verdant fields. To accomplish this dream, Chinese officials inaugurated a dizzying array of massive engineering projects: dams, reservoirs, irrigation systems. Every county in China was ordered to construct a water reservoir; most were dismal failures.[96] In the process, tens of millions of peasants were deployed in military fashion throughout China. Mass organized labor was approached with military precision and efficiency. In Xushui, local leader Zhang Guozhong approach irrigation projects like field campaigns, with conscripted workforces divided into battalions, companies, and platoons. The conscripts would sleep in makeshift barracks, far from the collective, commune, or village. This was soon replicated across the country.[97] Also, in some communes, "shock troops" were formed. These were special detachments of "exemplary" workers who would toil in the fields or on construction projects continuously for twenty-four hours at a time; they would sleep in makeshift barracks at the labor site, and eat in communal kitchens. [98]

Targets in water conservancy were calculated by the number of tons of earth that were moved in any given province; this magic number—entirely unrelated to the actual usefulness of any given project being undertaken—was then compared nationwide in a competitive spirit of emulation. These calculations in effect determined the political clout of a province; the tally-sheet was all that mattered.[99] That such calculations carried a human cost was well understood. In March 1958 Mao responded to the claim of a provincial official about an earth removal project: "Wu Zhipu claims he can move 30 billion cubic meters; I think 30,000 people will die. Zeng Xisheng has said that he will move 20 billion cubic meters, and I think that 20,000 people will die. Weiqing only promises 600 million cubic meters, maybe nobody will die."[100]

Most projects initiated during the Great Leap Forward were contradictory. The forcing of peasants to work backyard furnaces or to construct massive reservoirs and irrigation systems led to shortages of agricultural labor. One consequence was that in many locales, much-needed crops were left to rot in the fields. Likewise, the melting of shovels, hoes, and pickaxes—to make steel in the backyard furnaces—led to shortages of these implements both in the fields and at the construction sites. It was not uncommon that peasants would be forced to "deep plow" or to build irrigation canals with bare hands. In their totality, these programs and policies facilitated the political making of famine.

CHINA'S CULTURE OF IMPUNITY

Practices of moral exclusion and political violence are legitimated, justified, and sustained through complex imaginative geographies.[101] Both direct violence—the beatings, rapes, and tortures that occurred throughout the famine—and structural violence—the grain procurements, labor camps, and communes—are justified and spatialized through the forwarding of particular knowledges of "difference" between bodies in space. As Derek Gregory explains, imaginative geographies "are constructions that fold distance into difference through a series of spatializations. They work . . . by multiplying partitions and enclosures that serve to demonstrate 'the same' from 'the other.'"[102] As a spatial practice of moral exclusion, imaginative geographies construct gaps or differences between social groups by designating certain spaces as "ours" or "theirs." In China, "people's communes" became just such spaces, demarcating those whose lives were considered mere cogs in the greater socialist machinery.

When the plight of those who have been marginalized, both socially and spatially, is routinely neglected and ignored, we may properly speak of a pervasive "culture of impunity." Impunity, here, refers to the exemption from accountability, penalty, punishment, or legal sanctions for a crime. More broadly, we can identify a culture of impunity—the institutionalization of impunity—when torture and violence are overtly or tacitly condoned or unpunished as a result of amnesties, pardons, indifference, or simply "looking the other way."[103] It is a culture of impunity that permits widespread violence—such as that which occurred through China's famine—to continue unabated. It is also the *legacy* of a culture of impunity that tinges subsequent analyses of mass violence.

A culture of impunity serves to "blame the victims" of violence for their own plight. A growing number of studies have documented the tendency of both participants and bystanders of violence to blame the victims. This is seen at a variety of scales, ranging from an individual rape victim to the millions who died in the Holocaust.[104] To capture this concept, Melvin Lerner coined the term "just-world phenomenon" in reference to the practice of people believing that the world is just and that *other* people "get" what they deserve.[105] In other words, the social categorization of bodies into populations, and the concomitant dehumanization (or social death) of people not only recategorizes bodies into subhuman groups, it also carries with it an understanding that victims deserve or require their own victimization.[106] Waller elaborates:

> [A] strong belief in a just world is associated with [a] rigid application of social rules and belief in the importance of convention, as opposed to empathy and concern with human welfare. How do we explain, for instance, the plight of the poor and homeless? Those who display a strong belief in a just world are most likely to blame the poor and homeless for their own suffering. Similarly, some find it more comfortable to believe that battered spouses must have provoked their beatings; that sick people are responsible for their illness; that persons involved in a traffic accident must have been driving carelessly; that victims of theft surely brought it on themselves by not taking adequate security precautions.[107]

Or famine victims starved because of their own irrational behavior.

The basis of just-world beliefs is many and varied. For the most part, we are simply socialized into such a belief. Children, for example, are often taught that good behavior is rewarded, while bad behavior is punished. The family, consequently, is a major factor in the socialization of morals. In Maoist China, however, the family was steadily replaced not only by the Communist Party, but by Mao himself. Consequently, the promotion of just-world beliefs originated in, and was sustained by, the policies, practices, and pronouncements of Mao.

Moreover, the consequences of a just-world belief are foundational to exclusionary practices (e.g., the collectivization of households), the discipline of bodies, and the regulation of populations. Such a belief permits a level of indifference to—and thus a lack of engagement with—the plight of others. Peasants, farmers, so-called rightists, and especially those who were starving: all assumed the position of other within Maoist China and all were marginal—as *humans*—to the state. And when they could no longer produce, decimated by malnutrition, disease, and exhaustion, they became superfluous, expendable, not worthy to live. And when undernourishment and overwork reduced tens of millions of peasants to a condition where they were simply too enfeebled to work, it was Mao himself who proclaimed: "This won't do. Give them this amount [of food] and they don't work. Best halve the basic ration, so if they're hungry they have to try harder."[108]

Mao's knowledge/violence coupling set in place a pervasive culture of impunity, whereby the suffering and deaths of the peasantry were considered—by many officials—acceptable. It was more expedient to simply blame the victims for their own hardships—an expediency that continues to surround the legacy of the famine. Both then and now, two arguments have been forwarded to temper the violence surrounding the famine. The first can be easily dispatched. It was argued, and remains a point of contention, that peasants hid grain from the state. Increasing "record yields" translated into increased state procurement, as deliveries of grain to the state had to be made according to the official yields provided by local cadre. In 1958, for example, the actual crop yield was approximately 200 million tons; on the basis of claims made, however, party officials estimated that China had produced approximately 410 million tons. Significantly, if the state didn't receive its allotted amount, the implication was that farmers were hiding grain—in other words, stealing from the state.[109] Zhao Ziyang, secretary of Guangdong Province, for example, reported in January 1959 that many of the communes had hidden grain and hoarded cash; this is why local cadre were unable to meet the procurement targets established by the state. Thus, in response, Zhao launched an "anti-hiding" campaign that supposedly uncovered over one million tons of grain. This was reported to Mao.[110] In turn, when more and more stories of food shortages began to appear—after months of "record" yields—Mao concluded also that peasants were lying or, worse, were stealing and hiding food. He launched his own "anti-grain-concealment" campaign.[111] However, as Chang and Halliday conclude, "Mao knew perfectly well that the peasants had no food to hide. He had an efficient reporting system, and was on top of what was happening daily around the country."[112] These "anti-hiding" campaigns were rationalizations for the continuance of purges, broader efforts to remove what Mao (and others) considered to be disloyal subjects.

A second and more enduring example of blaming the victim lay in supposedly irrational behavior of the peasants. Chang and Wen, for example, conclude that "the precise cause [of the famine] was actually the failure in consumption efficiency." By this they mean "the behaviors and outcomes of irrational, wasteful, and inefficient consumption of food as a result of communal dining with a free food supply during the famine period."[113] While directing attention to the communes, Chang and Wen lay blame on the starving peasants. According to this argument, through irrational behavior, the peasants ate themselves to death in the early stages of the famine; and by 1959 and 1960, there was no food left.

The consumption efficiency argument hinges on two elements. On the one hand, this argument holds that peasants actually had sufficient access to nutrients in the canteens and, because these foodstuffs were "free," the peasants irrationally ate too much. However, as detailed earlier, within the canteens the withholding of food was actually a weapon. And even if food wasn't withheld for disciplinary purposes, most canteens throughout the famine were unable to meet the basic needs of the people. As Chang and Halliday document, peasants in China's collectives actually ate fewer calories than those received by the slave-laborers at Auschwitz.[114]

On the other hand, the argument holds that local officials were unaware that grain yields were in fact exaggerated. And in some instances, as Johnson notes, the "exaggeration of the 1958 grain output estimates clearly affected the course of the famine by misleading [some] high officials, local officials and individual farm families into believing that normal approaches to conserving food were unnecessary."[115] Some officials did apparently believe the inflated yields and, in the early months of 1958, did encourage the peasants to "eat their fill."[116] But how does the overall argument square with the knowledge that these officials were not simply "misled," for many were in fact the very same people who put forward the exaggerated claims? Were the actions of the peasants in 1958 irrational? Did they eat too much, thereby causing (or at least intensifying) the famine? To be sure, many farmers did slaughter and eat their farm animals rather than turn them over to the communes in 1958.[117] Likewise, in 1958 there was a deliberate campaign to take from farm people any food stocks that they might have.[118] If a family did not eat its reserves, it would surely lose those. And given the recent history of socialist practice, there was no reason for the families to believe that they would receive adequate provisions from the state. And indeed, as subsequent months demonstrated—despite claims of record surpluses—the farmers received less and less. Despite claims to the contrary, they did not receive back surplus grains that they themselves had produced. As Dikötter concludes:

> The idea that the state mistakenly took too much grain from the countryside because it assumed that the harvest was much bigger than it was is largely a myth. . . . [T]he party had evolved a set of political priorities which ignored the needs of the countryside. The leadership decided to increase grain exports to honor its foreign contracts and maintain its international reputation, to such an extent that a policy of "export above all else" was adopted in 1960. It chose to increase its foreign aid to its allies, shipping grain for free to countries like Albania. Priority was also given to the growing populations of Beijing, Tianjin, Shanghai and the province of Liaoning—the heartland of heavy industry—followed by the requirements of city people in general. The consequence of these political decisions was not only an increase in the proportion of procurements, but also an increase in the overall amount of grain handed over to the state.[119]

CONCLUSIONS

In 1961, as the extent of the famine was undeniable, loyal party officials made excuses. It was a series of climate-related catastrophes; it was the demands of the Soviet Union to pay back debts; it was the greedy overconsumption of peasants. Mostly, however, they deflected attention from Mao, choosing instead to shoulder the blame—and thus cement their own standing. Zhou Enlai, for example, assumed complete responsibility. Others were less specific in assignment responsibility, but still refused to isolate Mao. Liu Shaoqi, for example, asked: "Are the problems that have appeared over the past few years actually due to natural disasters or to shortcomings and errors we have made in our work?" He concluded that "the center is the principal culprit, we leaders are all responsible, let's not blame one department or one person alone."[120] Likewise, Li Fuchun concluded that "Chairman Mao's directives are entirely correct, but we, including the central organs, have made mistakes in executing them."[121]

Should blame be apportioned for the famine? Can we identify "intent" in this episode of mass violence? Many scholars, for example, do not attribute intentionality in Mao's actions, although some do acknowledge Mao's acceptance of the ensuing deaths. Chang and Halliday, for example, find that although "slaughter was not [Mao's] purpose with the Leap, he was more than ready for myriad deaths to result."[122] In other words, whereas Mao did not intend to kill 40 million of his citizens, he was prepared to let these people die—if their deaths served greater objectives. Valentino concludes also that "there is no evidence to suggest that Mao or other Chinese leaders deliberately engineered the famine." He continues that the primary causes of the famine "were a combination of deeply flawed agricultural policies and the attempt by local cadres to meet or exceed the regime's preposterously high grain production goals."[123] I agree that the famine does not conform to the United Nation's definition of genocide; Mao did not intentionally *kill* his subjects, as did the Nazi state. However, the pursuit for social justice should not be shackled by politically derived definitions that explicitly seek to remove atrocities and crimes against humanity from analysis. Mao's utopian worldview was to be achieved—could only be achieved—through widespread terror and violence. Mao and other party leaders were aware of the horrific conditions that engulfed China; they chose to ignore these and to let die millions of people.

Philosophers and medical ethicists have long contemplated the moral difference between killing and letting die.[124] As Green explains, "Killing and letting die are frequently taken to differ in that they respectively involve doing something to cause death and doing nothing to prevent death; and it is usually supposed that, if the distinction has moral significance, it is because it is worse to kill than to let die."[125] Significantly, many of these philosophical discussions focus on the moral difference between killing (murder) and starvation. James Rachels, for example, asks us to imagine a starving child in the room where you are now. Imagine also that you have a sandwich. Rachel suggests that most (hopefully all!) people would immediately offer the sandwich to the starving child. Moreover, very few people would expect anything in return, including "praise." Rachels however asks us to imagine what we would think of a person who did not offer that child the sandwich. Again, most people would consider that person to be a moral monster.[126]

Now, however, let us add a spatial component. What if that starving child was not in the same room as you, but lived many thousands of miles away, in another country perhaps. Rachels acknowledges that "the spatial location of the dying people hardly seems a relevant consideration," and yet, he concludes, "the location of the starving people does make a difference, psychologically, in how we feel."[127] We are thus confronted with the stark geography of morality. For many people—and governments—there is a spatiality to our actions. As both the physical and social distance between groups of people increases, it becomes easier to not include them in our moral universe. In chapter 1, I introduced the spatial logic of killing, noting that it becomes easier to kill as the distance between perpetrator and victim increases. Likewise, as the perceived social distance between the two people increases—as in many forms of structural violence—the act of killing becomes easier. So too is the act of letting die.

In Maoist China those 40 million people who perished were not in another country; and yet the Chinese state stood by and let them die. A pervasive knowledge/violence relationship ensured that those who were aware of the starvation did nothing. To be sure, if one spoke out, they themselves were subjected to humiliation, punishment, or execution. It is perhaps understandable, therefore, why so many stood by while millions died. This, however, must focus attention back on the Chinese state and to the practices forwarded by Mao and other top officials. The Chinese state, as a whole, *intentionally* did nothing to prevent these deaths; furthermore, the policies and programs forwarded by the Chinese state *explicitly* formed the conditions that permitted widespread starvation and mass death. From this perspective, I suggest that letting die is just as morally wrong as killing. I continue this theme in the next chapter.

NOTES

1. Jasper Becker, *Hungry Ghosts: Mao's Secret Famine* (New York: Henry Holt & Co., 1998); Gene Hsin Chang and Guanzhong James Wen, "Food Availability

versus Consumption Efficiency: Causes of the Chinese Famine," *China Economic Review* 9(1998): 157–66; D. Gale Johnson, "China's Great Famine: Introductory Remarks," *China Economic Review* 9(1998): 103–9; Carl Riskin, "Seven Questions about the Chinese Famine of 1959–61," *China Economic Review* 9(1998): 111–24; Justin Yifu Lin and Dennis Tao Yang, "On the Causes of China's Agricultural Crisis and the Great Leap Famine," *China Economic Review* 9(1998): 125–40; Dali L. Yang and Fubing Su, "The Politics of Famine and Reform in Rural China," *China Economic Review* 9(1998): 141–56; Frank Dikötter, *Mao's Great Famine: The History of China's Most Devastating Catastrophe, 1958–1962* (New York: Walker & Co., 2010); and Xin Meng, Nancy Qian, and Pierre Yared, "The Institutional Causes of China's Great Famine, 1959–1961," paper presented at the Centre for Economic Policy Research's Development Economics Symposium, June 2–3, 2010 (www.cepr .org/meets/wkcn/7/780/papers/Qianfinal.pdf).

2. Quoted in Dikötter, *Mao's Great Famine*, 88.

3. Robert Dirks, "Social Responses during Severe Food Shortages and Famine," *Current Anthropology* 21(1980): 21–44; at 23.

4. Dirks, "Social Responses," 23–24; see also G. B. Leyton, "Effects of Slow Starvation," *Lancet* 2(1946): 73–79; Ancel Keys, Josef Brozek, Austin Henschel, Olaf Mickelson, and Henry Taylor, *The Biology of Human Starvation* (Minneapolis: University of Minnesota Press, 195); and George Cahill, "Famine Symposium: Physiology of Acute Starvation in Man," *Ecology of Food and Nutrition* 6(1978): 221–30.

5. China is notably missing in Frank Chalk and Kurt Jonassohn, *The History and Sociology of Genocide: Analyses and Case Studies* (New Haven, CT: Yale University Press, 1990); and Samuel Totten and William S. Parsons (eds.), *Century of Genocide: Critical Essays and Eyewitness Accounts*, 3rd ed. (New York: Routledge, 2009).

6. Jung Chang and Jon Halliday, *Mao: The Unknown Story* (New York: Anchor Books, 2006).

7. Benjamin A. Valentino, *Final Solutions: Mass Killing and Genocide in the Twentieth Century* (Ithaca, NY: Cornell University Press, 2004), 117. Emphasis added.

8. The complete text is available at the United Nations website (http://www .un.org/millennium/law/iv-1.htm). Emphasis added.

9. Dali L. Yang and Fubing Su, "The Politics of Famine and Reform in Rural China," *China Economic Review* 9(1998): 141–56; at 141.

10. Dikötter, *Mao's Great Famine*, 13.

11. M. Sarup, *An Introductory Guide to Post-Structuralism and Postmodernism*, 2nd ed. (Athens: University of Georgia Press, 1993), 72.

12. Michel Foucault, *Discipline and Punish: The Birth of the Prison*, translated by Alan Sheridan (New York: Vintage Books, 1979), 26.

13. Michel Foucault, "Two Lectures," in *Power/Knowledge: Selected Interviews and Other Writings, 1972, 1977*, edited by C. Gordon (New York: Pantheon Books, 1980), 78–108; at 98.

14. Foucault, *Discipline and Punish*, 26.

15. Michel Foucault, *The History of Sexuality*, volume 1: *An Introduction*, translated by R. Hurley (New York: Vintage Books, 1990), 94–95.

16. Michel Foucault, "The Subject and Power," in *Power: Essential Works of Foucault, 1954–1984*, volume 3, edited by Paul Rabinow (New York: New Press, 2000), 326–48; at 340.

17. Philip Barker, *Michel Foucault: An Introduction* (Edinburgh: Edinburgh University Press, 1998), 38.

18. See, for example, Graham Hutchings, *Modern China: A Guide to a Century of Change* (Cambridge, MA: Harvard University Press, 2001); Colin Mackerras, *China in Transformation, 1900–1949*, 2nd ed. (Harlow, UK: Pearson, 2008).

19. Judith Shapiro, *Mao's War against Nature: Politics and the Environment in Revolutionary China* (Cambridge: Cambridge University Press, 2001), 2.

20. Shapiro, *Mao's War*, 2–3.

21. Quoted in Chang and Halliday, *Mao*, 423.

22. Shapiro, *Mao's War*, 27.

23. Shapiro, *Mao's War*, 27.

24. See, for example, Valentino, *Final Solutions*, 124; and Dikötter, *Mao's Great Famine*, 9.

25. Becker, *Hungry Ghosts*, 54.

26. Shapiro, *Mao's War*, 27.

27. Dikötter, *Mao's Great Famine*, 89; Becker, *Hungry Ghosts*, 199–200.

28. Dikötter, *Mao's Great Famine*, 288.

29. Dikötter, *Mao's Great Famine*, 94.

30. Dikötter, *Mao's Great Famine*, 22.

31. Dikötter, *Mao's Great Famine*, 102.

32. Shapiro, *Mao's War*, 4 and 15.

33. Shapiro, *Mao's War*, 4.

34. Becker, *Hungry Ghosts*, 57.

35. Shapiro, *Mao's War*, 4.

36. Shapiro, *Mao's War*, 4.

37. Shapiro, *Mao's War*, 3.

38. Dikötter, *Mao's Great Famine*, 9.

39. Dikötter, *Mao's Great Famine*, 21.

40. Dikötter, *Mao's Great Famine*, 36–37.

41. Chang and Halliday, *Mao*, 421.

42. Meng et al., "Institutional Causes," 5.

43. Dikötter, *Mao's Great Famine*, 76–79.

44. Dikötter, *Mao's Great Famine*, 106.

45. Dikötter, *Mao's Great Famine*, 127–28.

46. Becker, *Hungry Ghosts*, 64–65.

47. Becker, *Hungry Ghosts*, 68.

48. Dikötter, *Mao's Great Famine*, 39.

49. Becker, *Hungry Ghosts*, 67.

50. Dikötter, *Mao's Great Famine*, 40.

51. Becker, *Hungry Ghosts*, 111.

52. Valentino, *Final Solutions*, 125.

53. Chang and Halliday, *Mao*, 419.

54. Quoted in Dikötter, *Mao's Great Famine*, 129.

55. Becker, *Hungry Ghosts*, 93.

56. Chang and Halliday, *Mao*, 418.

57. Valentino, *Final Solutions*, 120.

58. Valentino, *Final Solutions*, 118.

59. Meng et al., "Institutional Causes," 6.

60. Meng et al., "Institutional Causes," 6.

61. Valentino, *Final Solutions*, 122.

62. Becker, *Hungry Ghosts*, 105.

63. Yang and Su, "Politics of Famine," 142.

64. Valentino, *Final Solutions*, 125.

65. Valentino, *Final Solutions*, 125.

66. Yang and Su, "Politics of Famine," 142.

67. Chang and Halliday, *Mao*, 427.

68. Becker, *Hungry Ghosts*, 142.

69. Chang and Halliday, *Mao*, 426.

70. Becker, *Hungry Ghosts*, 109–10.

71. Becker, *Hungry Ghosts*, 110.

72. Foucault, *Discipline and Punish*, 215.

73. R. Marsden, "A Political Technology of the Body: How Labour Is Organized into a Productive Force," *Critical Perspectives on Accounting* 9(1998): 99–136; at 120.

74. Foucault, *Discipline and Punish*, 25.

75. Foucault, *Discipline and Punish*, 25.

76. Foucault, *Discipline and Punish*, 138.

77. Chang and Halliday, *Mao*, 426.

78. Barker, *Foucault*, 37.

79. Chang and Halliday, *Mao*, 420.

80. Becker, *Hungry Ghosts*, 111.

81. See, for example, Chang and Halliday, *Mao*, 427.

82. Paul W. Kahn, *Sacred Violence: Torture, Terror, and Sovereignty* (Ann Arbor: University of Michigan Press, 2008).

83. Foucault, *Discipline and Punish*, 33.

84. Foucault, *Discipline and Punish*, 34.

85. See, for example, Dikötter, *Mao's Great Famine*, 292–305.

86. Dikötter, *Mao's Great Famine*, 57.

87. Dikötter, *Mao's Great Famine*, 57.

88. Dikötter, *Mao's Great Famine*, 58.

89. Chang and Halliday, *Mao*, 423.

90. Dikötter, *Mao's Great Famine*, 58.

91. Chang and Halliday, *Mao*, 424.

92. Dikötter, *Mao's Great Famine*, 61.

93. Chang and Halliday, *Mao*, 424. Chang and Halliday continue that the "good" steel was in fact produced by state-run steel mills, factories that were purchased through grain procurements.

94. Dikötter, *Mao's Great Famine*, 61.

95. Dikötter, *Mao's Great Famine*, 38–39.

96. Becker, *Hungry Ghosts*, 77.

97. Dikötter, *Mao's Great Famine*, 47.

98. Becker, *Hungry Ghosts*, 105. There were both male and female shock troops.

99. Dikötter, *Mao's Great Famine*, 29.

100. Quoted in Dikötter, *Mao's Great Famine*, 33.

101. Derek Gregory, *Geographical Imaginations* (Cambridge, MA: Blackwell, 1994); Derek Gregory, *The Colonial Present: Afghanistan, Palestine, Iraq* (Cambridge, MA: Blackwell, 2004).

102. Gregory, *The Colonial Present*, 17.

103. Susan Opotow, "Reconciliation in Time of Impunity: Challenges for Social Justice," *Social Justice Research* 14(2001): 149–70; at 158. See also James A. Tyner, *War, Violence, and Population: Making the Body Count* (New York: Guilford Press, 2009), 40–41.

104. James Waller, *Becoming Evil: How Ordinary People Commit Genocide and Mass Killing* (New York: Oxford University Press, 2002); see also James A. Tyner, *Space, Place, and Violence: Violence and the Embodied Geographies of Race, Sex, and Gender* (New York: Routledge, 2011).

105. Melvin J. Lerner, *The Belief in a Just World: A Fundamental Decision* (New York: Plenum Press, 1980).

106. Waller, *Becoming Evil*, 249.

107. Waller, *Becoming Evil*, 252.

108. Quoted in Chang and Halliday, *Mao*, 427.

109. Dikötter, *Mao's Great Famine*, 62.

110. Dikötter, *Mao's Great Famine*, 85.

111. Becker, *Hungry Ghosts*, 86.

112. Chang and Halliday, *Mao*, 420.

113. Chang and Wen, "Food Availability," 158.

114. Chang and Halliday, *Mao*, 429.

115. Johnson, "China's Great Famine," 104.

116. Dikötter, *Mao's Great Famine*, 41. See also Becker, *Hungry Ghosts*, 80.

117. Johnson, "China's Great Famine," 107.

118. Johnson, "China's Great Famine," 107.

119. Dikötter, *Mao's Great Famine*, 133.

120. Dikötter, *Mao's Great Famine*, 121.

121. Dikötter, *Mao's Great Famine*, 122.

122. Chang and Halliday, *Mao*, 430.

123. Valentino, *Final Solutions*, 125.

124. See, for example, James Rachels, "Killing and Starving to Death," *Philosophy* 54(1979): 159–71; O. H. Green, "Killing and Letting Die," *American Philosophical Quarterly* 17(1980): 195–204; Jonathan Glover, *Causing Death and Saving Lives* (New York: Penguin, 1990 [1977]); John Harris, *The Value of Life: An Introduction to Medical Ethics* (New York: Routledge, 1985); Will Cartwright, "Killing and Letting Die: A Defensible Distinction," *British Medical Bulletin* 52(1996): 354–61;

and Jeff McMahan, *The Ethics of Killing: Problems at the Margins of Life* (Oxford: Oxford University Press, 2002).

    125. Green, "Killing and Letting Die," 195.
    126. Rachels, "Killing and Starving to Death," 160.
    127. Rachels, "Killing and Starving to Death," 161.

Chapter Four

Normalizing the State: Cambodia

Between 1975 and 1979 Sarom Prak lived through the Cambodian genocide.[1] And during these interminably long years, Prak witnessed—and experienced—events that only Dante could dream: the disemboweling of men, women, and children; the beheading of people by machete; the use of pincers to cut women's nipples and breasts; the tearing out of fingernails. And for what purpose were people subjected to these mutilations? In his story, Prak provides no conclusion to account for the violence he saw. He can only conclude that he is a survivor, and that he wants us to "fully realize what happens when people slay other human beings."[2]

Cambodia's genocide was proportionately one of the worst instances of mass violence of the twentieth century. In just under four years, approximately one-quarter to one-third of the country's estimated eight million citizens died. They died from torture, execution, exhaustion, starvation, and disease. And they died because the newly installed state determined that some lives were worth living and thus had a place in the new society, whereas other lives did not.

The transformation of Cambodia into Democratic Kampuchea—as the country was renamed—was, from the perspective of the Communist Party of Kampuchea (CPK; also known as the Khmer Rouge), to be achieved by literally "smashing" all preexisting histories, geographies, and societies. The Khmer Rouge explicitly attempted to erase all that came before them, in an effort to establish a pure, communist-based utopia. Such was the underlying ideology that informed the destruction and construction of social relations and practices throughout the genocide. In this chapter I consider four interrelated components of the genocide: (1) the erasure of Cambodia and the writing of Democratic Kampuchea; (2) the pursuit of conformity through disciplinary and violent practices; (3) the knowing of a new Cambodia

111

through a reconstructed educational system; and (4) the disallowance of life through the practice of medicine. Combined, these components highlight the fact that corporeal differences were recognized—or, more properly, constructed—and thus killed or let die. I argue that the Khmer Rouge attempted to create an egalitarian society, a socialist utopia, through the annihilation of difference. Indeed, the executions, tortures, and starvations were justified and legitimated in conformance with a particular geographical imagination of a new state, a state that rhetorically was predicated on the ideas of justice and egalitarianism.

THE ERASURE OF CAMBODIA

On April 17, 1975, thousands of war-hardened Khmer Rouge soldiers poured into the streets of Phnom Penh, Cambodia's capital city. Coinciding with the cessation of the broader Indochina War that engulfed neighboring Vietnam and Laos, the Cambodian Civil War (1970–1975) was over. In five bloody years, the Khmer Rouge had defeated the US-supported Republican forces of the Lon Nol government. In the process, tens of thousands of people had died; many hundreds of thousands found themselves refugees in their own land.

Much like the ascension of Adolf Hitler and the Nazi Party in Germany, the Khmer Rouge victory was far from inevitable. Throughout most of rural Cambodia in the 1950s and 1960s, preconditions for revolution were difficult to discern.[3] For many decades a colony of France, Cambodia achieved independence in 1954. Earlier "anticolonial" demands were thus superfluous. And while some discontent was directed against Cambodia's leader, Prince Sihanouk, most Cambodians were relatively content. As Chandler explains, most Cambodians—Khmer—"were reluctant to become involved in rebellious politics after Cambodia's independence had been won." Moreover, the individualism, conservatism, Buddhist ethics, and the fact that nearly all of the Cambodian peasantry actually owned their land made them unlikely candidates for communist recruitment.[4]

The communist movement in Cambodia owed its existence to the emergence of communism in Vietnam; and Vietnam's movement was, in part, facilitated by both the activities of Mao Zedong in China and Joseph Stalin in the Soviet Union. In fact, the first communist party to emerge in Cambodia during the 1930s was a cell of Ho Chi Minh's Vietnam-based Indochina Communist Party. Tellingly, this early "party" was composed primarily of ethnic Vietnamese laborers, with only a few Khmer.[5] It was subsequently reorganized in 1951 and renamed as the Khmer People's Revolutionary Party (KPRP).

Through the 1950s the incipient communist movement in Cambodia was mostly rural based, and heavily influenced by the Vietnamese. However, in the 1960s a splinter group emerged within the party, one that was decidedly more urban based, radical, and staunchly anti-Vietnamese. This shift was precipitated by the actions of a small group of Paris-educated Khmer who, within a few years, would take control of the KPRP—to be renamed the Communist Party of Kampuchea (CPK). In time, the CPK—known as the Khmer Rouge, a term popularized by Sihanouk—would be dominated by a single individual who modeled himself after Mao Zedong. The man's name was Saloth Sar, better known as Pol Pot. He would be joined by, among others, Nuon Chea, deputy secretary-general and president of the Representative Assembly of Democratic Kampuchea; Ieng Sary, deputy prime minister in charge of foreign affairs; Son Sen, deputy prime minister for defense and security; Khieu Samphan, president of Democratic Kampuchea; Ieng Thirith, wife of Ieng Sary and minister of social affairs and action; and Yun Yat, wife of Son Sen and minister of culture.[6]

The victory of the Khmer Rouge in 1975 would mark the termination of years of military conflict but not the end of widespread violence. In the weeks and months that followed, the cities and towns of Cambodia were evacuated, their inhabitants forced onto agricultural collectives and labor camps distributed throughout the countryside. Hospitals, factories, and schools were closed; money and wages abolished; monasteries emptied.[7] All existing societal relations and infrastructures were to be "smashed": health, education, commerce, religion, family. Such a focus on the destruction meted out by the Khmer Rouge, however, tells only half the story. What is less discussed in many accounts of the Cambodian genocide is that the Khmer Rouge also intended to *construct* an entirely new state and society.[8] The objective of Pol Pot, Ieng Sary, Khieu Samphan, and other top-ranking leaders of the CPK was to make an entirely new, modern, and productive communal society. In short, the goal of the Khmer Rouge was in fact twofold: to first erase all vestiges of the previous society and, second, to erect an entirely new, socialist society. It was with this understanding that Cambodia ceased to exist, replaced by Democratic Kampuchea. And it is here that the Cambodian genocide differs markedly from both the Holocaust and even Mao Zedong's Great Leap Forward. Whereas in the former, the Nazi state eliminated those individuals deemed unworthy of life, and built new monuments and buildings testifying to its glory, there was not a concerted effort to literally demolish all vestiges of earlier societies. Likewise, in China, party leaders "let die" millions of their citizens and attempted to reorient social life, but there was not an effort to wipe clean, so to speak, all that came before 1949. In Cambodia, the Khmer Rouge were clear on their intent, as reflected by the slogan "Destroy the old order, replace it with the new."[9] Significantly, Henri Lefebvre writes: "A revolution that does not produce a new space has not realized its full potential."[10] The resonance of Lefebvre's statement with that of the Khmer Rouge should not go unnoticed, for it strikes at the core of our understanding of the CPK's geographical imagination.

In chapter 1, I indicate that a geographic approach to genocide asks: Who, or which group, is granted or denied access to certain places? What activities are deemed appropriate or not? What relations of power are maintained when "place" is invoked? And who has the authority, the ability, to define (and enforce) those places? I argue that the processes leading to moral inclusion or exclusion have a *geographic* component, one that is infused with power—a power we might phrase *a will to space*. In Nazi Germany, this materialized as a quest for *lebensraum*; in Maoist China, it appeared as a communal utopia. Cambodia was different from its predecessors. Consider, firstly, that geography, strictly defined, is about the writing of space: geo (earth) and graph (to write). Geography as a verb is thus the writing, the building, the making of places. Now, if one accepts this premise, one must also consider the obverse, namely, the idea of antigeographies and the *erasure* of space. And here we begin to understand that the transformation of Cambodia envisioned by the Khmer Rouge was to be total. For the revolution to succeed, from the standpoint of the Khmer Rouge, all of Cambodia had to be first erased before the CPK would begin to write anew.

Nearly one century ago the geographer Derwent Whittlesey proposed the concept of "sequent occupance." He suggested that it is possible to study a landscape and to ascertain previous settlement patterns—in other words, a sequence of different forms of occupance "written" on the land. Whittlesey, for example, studied a small district in northern New England, in the United States, and identified a series of settlements, including those of livestock pasturage, farming, and so on.[11] This notion of landscape transition has been retained in subsequent theorizations of space, most notably in the writings of Lefebvre. As detailed in *The Production of Space*, Lefebvre provides a historical account of the production of space through a spatialization of Marx's modes of production. Lefebvre postulates that every "society to which history gave rise within a framework of a particular mode of production, and which bore the stamp of that mode of production's inherent characteristics, shaped its own space."[12] Hence, dominant economic systems, such as feudalism or merchant capitalism, will be manifest on the landscape. However, since "each mode of production has its own particular space, the shift from one mode of production to another must entail the production of a new space."[13] Landscapes of feudalism give way to landscapes of merchant capitalism; these give way to landscapes of industrial capitalism. In an ongoing dialectical process, as revolutions transform one economic system to another, the landscape reflects the accumulated burden of previous revolutions, societies, and modes of production. And even a cursory observation of most any European landscape reveals this process: from the enduring legacy of agricultural field patterns, medieval roads, and centuries-old castles and churches, to the smokestacks and canals of the industrial revolution. The Cambodian landscape of 1975 would have revealed vestiges of an indigenous precapitalist Khmer society, a French colonial mercantilist presence, and a newly applied veneer of Western industrial capitalism.

The Khmer Rouge's will to space was predicated not on an attempt to write "on top" of these previous societal influences, but instead to start clean: the "blank sheet of paper." The Khmer Rouge leadership was not content with retaining, and building on, past inscriptions of previous modes of production. Rather, these leaders attempted to replace what they saw as impediments to national autonomy and social justice with revolutionary energy and incentives.[14] In their attempt to create—not *re*-create—a utopian society, the leadership of the Khmer Rouge embarked on a massive program of social and spatial engineering. However, the CPK had a very precarious hold on power—a political situation that would have remarkable consequences. Specifically, the CPK understood that they would have to "build socialism." This concept is derived from the theorizing of post-Leninist-Marxian economists who realized that revolutions do not always take place after the final phase of capitalist evolution and, therefore, that the concentration of productive forces would have to be achieved under revolutionary control.[15] Consequently, in order to achieve socialism and proceed towards a communist utopia, transition phrases should be identified and implemented. Thion argues, however, that the CPK grossly misunderstood the idea of "transitional" phases; for the Khmer Rouge, these were anathema—transitional phases were considered "reformist" or "revisionist." Instead, the Khmer leadership believed that they could, similar to Mao's China, jump over the transition phases—to bring about a "super great leap"—and achieve instant socialism.[16] This line of reasoning would have grave consequences.

The transformation of Cambodia into Democratic Kampuchea, from the perspective of the CPK, was to "smash" all preexisting histories, geographies, and societies. Consequently, the Khmer Rouge explicitly attempted to "unwrite"—to erase—both history and geography to create (in their geographical imagination) a pure, communist-based utopia. The Khmer Rouge did not try to turn back the pages of time to an earlier era—as a common myth of the genocide holds—but instead tried to rush forward at a dizzying pace regardless of the consequences.[17] This erasure of space is most clearly evident in the Khmer Rouge's decision to evacuate (and destroy) urban areas.

To achieve their political and economic goals, the Khmer Rouge drew on the radical extremes of Mao's Great Leap Forward: to bring about a complete and rapid transformation of Cambodia, to establish a self-sufficient state free from foreign intervention. To accomplish their "Super Great Leap Forward," however, the Khmer Rouge—as I argue—believed it necessary to erase any barriers to the revolution. The transformation of Cambodia was predicated on the Khmer Rouge's ability to literally crush or smash all elements of earlier societies. Cities, as the embodiment of foreign influence, were targeted first. As early as 1971 Khmer Rouge forces began systematically to burn rural villages and hamlets that fell under their control; their intent was to force peasants into new communal agricultural collectives.[18] In September 1973, for example, Khmer Rouge forces seized half of Kampong Cham City and "took fifteen thousand townspeople into the countryside with them."[19] Also, in March 1974, the Khmer Rouge army captured the city of Oudong and led more than 20,000 people into the countryside. In the process, the Khmer Rouge killed all schoolteachers and government officials, and deliberately razed the town.[20]

Such physical destruction was to provide a blank page, an opportunity to radically build a new society and to mark a clear and permanent break from the past.[21] The destruction of urban areas thus dovetailed with the Khmer Rouge's attempt to "build socialism." In their quest to make a Super Great Leap Forward into communism, the Khmer Rouge justified their brutal practices of forced evacuations and resettlement schemes. And having emptied the cities and begun the task of forming collectives, the Khmer Rouge turned to molding a new population.

MAKING POLITICAL SUBJECTS

In conforming with the ideology of erasing all vestiges of previous societies, the Khmer Rouge viewed certain populations and certain "bodies" as "good" or "bad," "pure" or "impure"; accordingly, these valuations of lives would directly relate to how people were to be used—or eliminated.[22]

A traditional saying in Cambodia holds that "clay is molded while it is soft."[23] According to Henri Locard, this slogan was often used to signify that only young children could be selected by the CPK to become loyal servants of Angkar. This idea, in fact, was developed by Pol Pot, who said of the young: "Those, among our comrades, who are young, must make a great effort to re-educate themselves. They must never allow themselves to lose sight of this goal. You have to be, and remain, faithful to the revolution. People age quickly. Being young, you are at the most receptive age, and capable to assimilate what the revolution stands for, better than anyone else."[24]

In Democratic Kampuchea, the Khmer Rouge routinely, and somewhat arbitrarily, constructed social groups: "new people" and "old people," capitalists, landlords, intellectuals—a subject I expound on later. Certain bodies were constructed as *different.* Difference within Democratic Kampuchea, however, was the antithesis of *equality.* Consequently, the Khmer Rouge attempted to achieve a complete state of conformity. All people were to be alike, all were to be identical. Any deviation from the general party line—any selfish or individual act—could result in severe punishment and probable death.[25] Thus, as the Khmer Rouge attempted to promote a singular sameness of people, one based on an idealized geographical vision of the peasantry, all people were to dress and groom alike. Men and women were to wear traditional black peasant garb; hairstyles were also regimented.

Through the erasure of traditional social relations, the Khmer Rouge attempted to regulate its population through the control of reproduction. Weddings were to be determined by the state; all other forms of affection were prohibited. According to Sarom Prak, "Flirtations, adultery, and love affairs were reasons for execution."[26] Traditionally, marriages were arranged in the interest of the family, as well as the potential bride and groom. Decisions were family affairs, and involved the consultation and permission of both

Figure 4.1. Staff of Tuol Sleng Security Center dining with their families. Kaing Guek Eav, alias Duch, is standing to the side. *Source:* Documentation Center of Cambodia

Figure 4.2. Two female Khmer Rouge cadre stand at attention near a work camp. *Source:* Documentation Center of Cambodia

parents and extended family members.[27] Under the Khmer Rouge, however, marriage was maintained as a state institution. Local party officials replaced parents in the negotiation of marriages. Individuals (in general, but not always) were denied the right to choose their own spouse; parents and other family members were forbidden to be involved in the decision.[28] In practice, marriages were determined by the social groupings constructed by the Khmer

Rouge. People were permitted to marry only within their own "class" or "category": soldiers could marry soldiers, for example.

Symbolically, the wedding did not represent the union of a couple or the beginnings of a family; rather, the marriage spectacle—for "mass" weddings were common—actually drew attention away from the significance of the individual and the family and toward their obligation to the state. Weddings in Democratic Kampuchea were not individual events, but communal tasks to be performed. Couples, perhaps fifty at a time, would line up in rows, as rank-and-file soldiers might, with females on one side, and males on the other. During the ceremony, couples—which might have only recently been introduced—would hold hands and pledge their loyalty to Angkar. Marriages were to serve reproductive purposes. As Peg LeVine writes, these conscripted weddings became part of one's duty to country to propagate (literally) the state.[29] However, violence was an ever-present threat to the practice of marriage under the Khmer Rouge. Prak explains that the "forced marriages were a good way for some of the people who had power—like the soldiers, the chiefs of villages and districts—to molest young girls until they got pregnant." He notes, also, that many of these girls were killed if they refused to be married.[30]

Beyond these, the Khmer Rouge attempted to implement a complete and total subjugation of the body and mind. All aspects of individualism were to be destroyed. Those who did not conform, those who stood apart, were considered traitors to the revolution, the party, and the state. Citizens of Democratic Kampuchea were to lead simple rural lives—as all cities were to be abolished. With military precision and discipline, they were to labor in the fields, producing rice for the state. And drawing inspiration from China, they set about massive engineering programs: irrigation systems, dams, reservoirs. Hundreds of thousands of workers, increasingly weakened by starvation and exhaustion, labored across the countryside in a monumental effort to achieve a Super Great Leap Forward.

The Khmer Rouge sought to *normalize* Democratic Kampuchea. As Foucault explains, "When you have a normalizing society, you have a power which [facilitates] an indispensable precondition that allows someone to be killed, that allows others to be killed."[31] And within Democratic Kampuchea, many "types" of people were to be normalized, classified, calculated, and evaluated. Such conformity required a total and complete spatial system of discipline. As discussed in earlier chapters, discipline is a form of power, a set of techniques, instruments, procedures, and applications.[32] And similar to Maoist China, discipline for the Khmer Rouge was also predicated on forms of enclosure. Throughout Democratic Kampuchea, following the evacuation of the cities, millions of peasants were forced into collectives.

Although reality often belied rhetoric, the Khmer Rouge leadership approached the task with pragmatism. Indeed, the Khmer Rouge acted much as "scientists" or "regional planners" when it came to the spatial organization of society. Consider, for example, the "excerpted report" of a speech delivered by a "comrade representing the Party Organization" at an Assembly of CPK members in 1976 in the "Western" zone of the country.[33] The speaker, most likely Pol Pot, discussed at length the "geography" of Democratic Kampuchea. Throughout the report, the speaker is dispassionate but optimistic, as he proceeds through an evaluation of regional opportunities. A series of rhetorical questions are asked, followed by straightforward, though often rambling, responses. The speaker, for example, identifies that "the geography of the Zone includes many mountains and forests, few plains and ricelands." He then asks: "Can we make a Super Great Leap Forward or not?" Answering in the affirmative, the speaker explains: "We must look into this problem and resolve it."[34] This resolution is predicated on a specific, calculated management of the human and physical geography of Democratic Kampuchea. The speaker explains:

> To break even means that [workers in] the No. 1 system get three cans (of rice per day); and No. 2, two-and-a-half cans; in No. 3, two cans; and No. 4, one-and-a-half cans. But we do not want only this much. We want twice as much as this in order to have capital to build up the Zone and the country. We must have enough to eat, that is 13 *thang* per person per year. If the Zone has 600,000 people, they must eat 150,000 tons of paddy. But we want more than this in order to locate much additional oil, to get ever more rice mills, threshing machines, water pumps, and means of transportation, but as an auxiliary manual force and to give strength to our forces of production. So we must not get just 150,000 tons of paddy. We must get 300,000, 400,000, 500,000 tons just to break even and be able to build socialism and completely get away from the former system and out of the former period.[35]

Within this passage one sees clearly the attempt to calculate production levels for the country, and how Democratic Kampuchea would be able to make a Super Great Leap Forward based on agricultural surpluses. Party leaders—most notably Pol Pot—drew on the radical ideas of Mao's Great Leap Forward. Economic priorities were based on agricultural production. The intent was to increase the national production of rice and, subsequently, export the surpluses in exchange for hard currency. These monies would then be used to finance heavy industry.[36] In so doing, the CPK sought to transform Cambodia into a modern, communal utopia.

The speaker considers the problems of regional diversity for economic productivity. Thus, the speaker explains that "in Region A there are two areas: the upper and the lower areas. The lower area is a good area with very great

Figure 4.3. **Transporting dirt on an irrigation project.** *Source:* Documentation Center of Cambodia

potential. It can grow both rice and vegetables and also has fish. The same goes for Region B. Its rich area is district C. If we produce only two tons of broadcast rice . . . this is not an appropriate amount. We must attack so as to get eight tons. For example, on 5,000 hectares of land, the produce is up to 40,000 tons—20,000 to support the 80,000 people, that is, to support all the people in Region B, and there are still 20,000 tons of paddy left. So if we attack on target, just in the one district C, we can feed the whole Region and with half still left over."[37]

To be sure, the production targets—although "attacked" with forced labor—were not met; starvation was widespread throughout the country. My point though in providing this discussion is to highlight the "scientific" way in which the Khmer Rouge operated. Here, their spatial planning conforms more to the brutally deliberate organization of the Nazi state than it does to the slavish fervor that characterized China's Great Leap Forward. Such a rationale underlay the Khmer Rouge's approach to collectives. Again, reality failed to match rhetoric, but the calculations are themselves telling. As detailed in the CPK's "Four-Year Plan," introduced during the summer of 1976, there was a set of procedures—and requirements—that were to accompany the construction of collectives.[38] In principle, collectives were to have "neat, clean, and proper hous[es] for each family"; "watering places for people and animals"; "places for animals to live in"; "hygienic toilets"; "sheds for fertilizers"; "carpentry workshops"; "kitchens and eating houses"; "schools and meeting places"; "medical clinics"; "vegetable gardens"; "barbers and hairdressers"; "warehouses" and places for "tailoring and darning."[39]

Needless to say, collectives throughout Democratic Kampuchea were far from these ideal plans. Instead, and more in line with those of Maoist China, collectives were places of rigid discipline, harsh punishments, and violence. Prior to the revolution, the extended family was the center of Cambodian economic and rural life. Families worked together as an economic unit responsible for household production and consumption. That said, households in prerevolutionary Cambodia exhibited a variety of forms. While many families lived in nuclear arrangements, others lived in "stem" arrangements, wherein one child would remain at home following marriage, bringing his or her spouses into the household. These arrangements could be either patrilocal or matrilocal. Still other families were considered extended, as when orphaned nephews or nieces were taken into the home.[40]

Traditional Khmer households were subsistence based. The mainstay of most farmers was rice, although other foods and household items were also produced at home. Families commonly owned fruit trees and palms; livestock was raised and vegetables and herbs were cultivated in family gardens. It was not uncommon for households to raise fish in rice paddies during the rainy season. However, despite this self-sufficiency, villages did engage in trade; many villages were closely tied to regional and national market economies. Households, for example, were linked to the cash economy to pay taxes or to pay off debts. Monies were also required to pay for weddings or funerals. Consequently, peasants would sell surplus produce or provided labor in exchange for wage payment.[41]

Under the Khmer Rouge, the ownership of production and consumption became increasingly communal. Cooperatives were established to cultivate specified crops, including rice, vegetables, fruits, and nonfood crops such as cotton, rubber, and jute. The collection and distribution of produce and goods was centralized, and large-scale collectives, coupled with irrigation projects and the development of (poorly planned and constructed) reservoirs and dams, were designed to foster greater farm yields. Furthermore, and consistent with their "scientific" attempt to "modernize" Cambodia through socialist practice, the Khmer Rouge initiated a spatial practice to "rationalize" the arrangement of rice paddies. Previously, rice paddies were arranged haphazardly, in parcels of varying shapes and sizes, the result being a pastiche of land-use patterns. Under the Khmer Rouge these paddies were reorganized into regular, quadrangular plots—supposedly a more efficient spatial arrangement of farming practices.[42]

Ranging in size from several hundred persons to several thousand, these communal farms were the site of concentrated labor activities that were meant to achieve economic self-sufficiency for the Khmer Rouge. Within the collectives, moreover, the spatial separation of daily life was augmented with respect to social and economic relations. In the process, the spatiality of daily life within the country served to break the traditional familial bonds and redirect loyalties to the state. As explained by Teeda Butt Mam, who was fifteen years old when Cambodia ceased to exist, the Khmer Rouge "kept moving us around, from the fields into the woods. They purposely did this to disorient us so they could have complete control. They did it to get rid of the 'useless people.' Those who were too old or too weak to work. Those who did not produce their quota."[43] Such was the calculation of life under the Khmer Rouge.

According to Michel Foucault, "Each individual has his [*sic*] own place; and each place its individual." He explains that "disciplinary space tends to be divided into as many sections as there are bodies or elements to be distributed. One must eliminate the effects of imprecise distributions, the uncontrolled disappearance of individuals, their diffuse circulation, the unusable and dangerous coagulation," Combined, "enclosure" and "partitioning" function as tactics of "anti-desertion, anti-vagabondage, anti-concentration." The aim is to "establish presences and absences, to know where and how to locate individuals." These are, in effect, procedures "aimed at knowing, mastering and using."[44]

Within the work brigades and collectives established by the Khmer Rouge, individual bodies were under constant watch; their activities were monitored and, at times, recorded. Any movements, any attitudes, any thoughts that were "out of order," that did not conform, were subject to immediate and brutal punishment. Bodies were to remain portioned, each sequestered into their own spaces. Here is how the Four-Year Plan described the distribution of food rations: "Organize, nominate and administer people to take responsibility for cooking tasty and high-quality food and desserts; i.e., there must be a separate group, not people taking turns, who are responsible for cooking and making desserts and consider it a revolutionary duty."[45] The plan similarly provided a schedule for the people's "working and resting regime": three days of rest per month; one rest day in every ten; between ten and fifteen days, according to remoteness of location, for rest, visiting, and study each year; and two months' rest for pregnancy and confinement. The document continues: "Resting time at home is nominated and arranged as time for tending small gardens, cleaning up, hygiene, and light study of culture and politics."[46]

Such ideals were not reflective of the actual conditions. Of their work on the communal farms, Quinn quotes a young Khmer Rouge cadre:

From now on if the people want to eat, they should go out and work in the rice paddies. They should learn that their lives depend on a grain of rice. Plowing the soil, planting and harvesting rice will teach them the real value of things.[47]

Another survivor, Roeun Sam was fourteen years old when the Khmer Rouge came to power.[48] She recalls:

> They put me with others my age and had me work in the field to watch the cows. Every day I watched the cows, and after I fed them at night I went to the place where the children laid on the ground to sleep. We didn't have a roof, wall, or bed. We only ate one meal a day, at lunch. Angka measured each serving, only about a cup and a half, which was mostly broth and maybe two tablespoons of rice.[49]

In her narrative, Sam describes the partitioning of Khmer society, of how different people—classified by age and sex—were assigned to different and specific places. This conformed to the overall intention of the Khmer Rouge leadership—that all aspects of quotidian life be regimented and geographically controlled to better serve the revolution and the state. The Khmer people were divided into work "forces" or "teams" based on age and sex. Adults, those aged between fourteen and fifty, were forced into mobile work brigades termed *kong chalet*. Men were assigned to *kong boroh* and women to *kong neary*. The heaviest work was performed by these two groups: plowing fields, planting and transplanting rice, harvesting, and the construction of irrigation canals and reservoirs.[50]

Figure 4.4.   Women working on an irrigation project. *Source:* Documentation Center of Cambodia

Older members of Democratic Kampuchea were grouped into work teams known as *senah chun*. Again, these were divided by sex, with men belonging to *senah chun boroh* and women belonging to *senah chun neary*. Those in *senah chun* usually performed lighter work: sewing, tending gardens, collecting wood, caring for children. Unlike the *kong chalet*—who often traveled long distances to particular work sites—those members of the *sehan chun* remained close to the collective. Lastly, the children of Democratic Kampuchea were also organized into sex-segregated work groups called *kong komar*. These members had the lightest work, such as tending water buffalo, collecting fuel wood, or gathering cow dung for fertilizer.[51]

Collectivization, and the attendant communal dining, radically altered traditional Khmer society. As work and dining were increasingly collectivized between 1976 and 1977, people associated more with their work teams than with their families. Taking a page from Maoist China, these sociospatial practices were to both severe familial bonds and provide for better supervision. As May Ebihara explains, the "imposition of communal dining halls was not simply a means whereby the state controlled the distribution; it further demonstrated that the work team or cooperative had superseded the family as the basic social unit in Democratic Kampuchea."[52] Likewise, Kalyanee Mam writes that "collective dining was enforced . . . because the regime feared that

Figure 4.5. Men working on an irrigation project along the Chinith River, Kampong Thom Province, 1976. The picture was taken by a Chinese photographer. *Source:* Documentation Center of Cambodia

Figure 4.6. Construction on an irrigation project. *Source:* Documentation Center of Cambodia

allowing families to produce their own food would encourage family interest and distract loyalty from *Angka*. As with other policies implemented by the regime, the purpose of collectivizing food and property was to eliminate individual dependency on the family and [to] force individuals to project this dependency towards the organization."[53]

Violence was also used as a means of ensuring discipline. Within Democratic Kampuchea, the ritualized practice of torture was frequently conducted at reeducation or self-criticism sessions. According to Henri Locard, self-criticism and mutual criticism constituted basic mental exercises.[54] Children and adults were required to attend nightly meetings—after working long hours in the fields—and to announce their shortcomings for all to hear. Throughout the sessions, the Khmer Rouge leaders would admonish the people: "If you committed an error, criticize yourself first, then punish yourself." The people were to "learn from model comrades, learn from . . . magnificent people, and forever learn from valuable documents, concentrate . . . to increase [understanding of] the discipline of *Angkar* and the theories of Marx-Lenin."[55] As localized show trials, these every-day sessions were to encourage public denunciations of malingerers, to expose "enemies" of the revolution, and to confess to deviant behavior. Discipline and the threat of violence became banal, ordinary, a feature of daily life. Locard continues: "Accusations, arising almost always from commune leaders, invariably led to arrest and then to liquidation [execution]. How often in public did people confess their wrongs, errors, flaws, or denounce their neighbors, almost always under threat and in fear for their own safety? People did their best to be invisible, lower their heads, and keep their opinions to themselves."[56]

The public display of torture and execution served to reaffirm the authority of the government. Ngor explains: "The soldiers took captives morning and afternoon. Instead of marching them away immediately, the soldiers made public examples of them, tying them to trees and shouting to anyone who would listen what they had done wrong."[57] Other practices were "hidden" in plain sight. It was common for people to simply disappear, taken to killing sites. Ouk Villa, for example, describes waking up one night and, as a curious child, running off after hearing some commotion: "I ran through the rice fields and the bushes. . . . Suddenly I saw three men who were tied up and being led by a militiaman to another small bush. From a distance I saw the militiaman force the men to kneel down on the edge of a big pit. A minute later the men were clubbed to death with a hoe. I could see this clearly in the moonlit night."[58] Roeun Sam describes a similar experience. She remembers that one day some of the cows she was tending had wandered away. She writes: "I smelled something like a dead animal. My cows were running toward the smell, and I followed them. By the time I got there the cows were licking the dead body's clothes. . . . It was a human body that had just been killed. You could see her long black hair and the string around her hands. I looked around and saw people who had been shot and their heads were smashed in. There were at least one hundred people dead."[59] These places where Sam and Ouk had discovered bodies were in Sam's haunting words, where the Khmer Rouge took people to kill.[60]

The Khmer Rouge's terror state was also supported by an extensive prison system. According to Khamboly Dy, security centers (prisons) were organized into five levels.[61] These were used for detention, interrogation, and execution. At the lowest level of organization (and size) were regional, district, and subdistrict centers. Most prisoners held at these centers were accused of stealing, desertion, or speaking ill of the government. Punishments were "usually" not as severe and it was not uncommon for prisoners to be released and sent back to labor camps. However, as the memoir of Ngor makes clear, the spectacle of violence was ever present. Ngor describes his experience of being held at a local prison and witnessing the murder of a pregnant woman.[62]

Beyond these local prisons were zone level security centers and the central security centers—such as the infamous S-21. These centers would often hold a thousand or more prisoners, primarily those Khmer Rouge officials, soldiers, and their families who had been accused of committing major offenses, such as treason. Torture was the rule of the day at these centers; execution a foregone conclusion. At S-21, located in the heart of Phnom Penh, over

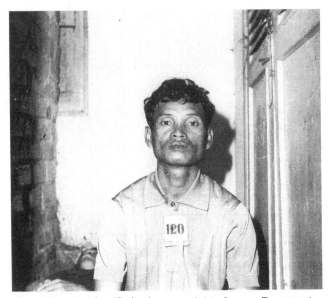

Figure 4.7.   Unidentified prisoner at S-21. *Source:* Documentation Center of Cambodia

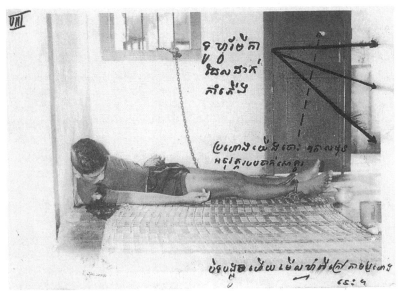

Figure 4.8.   Unidentified prisoner who committed suicide at S-21. *Source:* Documentation Center of Cambodia

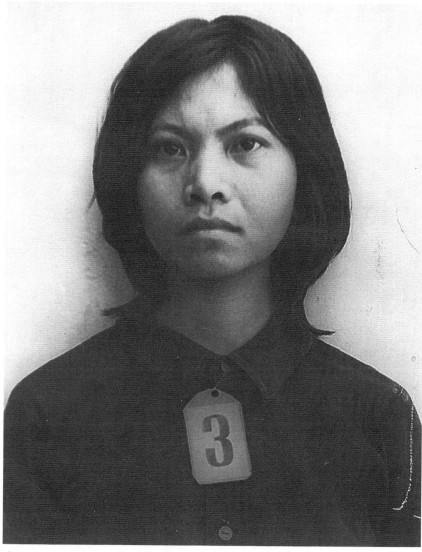

Figure 4.9. Mug shot of Hout Bophana, arrested on October 10, 1976. She was interrogated and tortured for five months before being executed in March 1977. *Source: Documentation Center of Cambodia*

14,000 men, women, and children were detained and tortured before being taken to the killing fields at Choeung Ek, where their lives would end not by bullet, but by a club to the head. Of those prisoners who entered S-21, only a fraction are known to have survived.[63]

IMAGINING GEOGRAPHY THROUGH EDUCATION

Peter Taylor suggested that "God invented war to teach Americans geogra-phy."[64] Taylor's statement, of course, should be taken with a grain of salt. The point he is making is that there is a long-standing relationship between geog-raphy as an academic body of knowledge and military activities.[65] Moreover, and this strikes at the core of Taylor's statement, war promotes geographic *awareness*. Susan Schulten, for example, notes that following certain pivotal moments during World War II—the German invasion of Poland, the assault on Normandy—Americans bought in a matter of hours what in peacetime would have been a year's supply of maps and atlases. She continues that this nationwide attention to maps brought the farthest reaches of war into every-day conversation, and demonstrated the powerful relationship between war and geography.[66] My point is not to suggest that geography lies at the heart of war, but rather to reinforce the idea that geographical *imaginations* do play a significant role in the justification and legitimating of mass violence. Geog-raphy, as Schulten explains, is a way of distinguishing "here" from "there," without which little sense can be made of human experience.[67]

As previous sections indicate, the Khmer Rouge sought to erase Cambodia; in its stead was to be a new, better, forward-looking country called Democratic Kampuchea. Moreover, this new state was to be populated by a disciplined citizenry that exhibited an unquestioned sameness. But how was one to know one's people and country? The answer is quite simple: teach the children.

The Khmer Rouge understood the importance of education in their post-conflict construction of Democratic Kampuchea. Indeed, education was vital to their revolutionary project in that it would provide support and legitimacy for associated political and economic programs. When the Khmer Rouge stood victorious on the streets of Phnom Penh they constituted neither a cen-tralized, efficient political party nor military force. Having achieved military victory, the Khmer leadership understood that they would have to centralize power and "build socialism."

With such a tentative hold on the populace—and its own political power—the Khmer Rouge leadership sought to solidify their position through various means. On the one hand, the Khmer Rouge utilized a practice of state terror. Within Democratic Kampuchea, for example, the public display of torture and execution served to reify the authority of the Khmer Rouge. Moreover, the systematic violence and the killing of its own populace were understood by the Khmer Rouge as a prelude to the construction of a moral and properly ordered postwar society.

On the other hand, the Khmer Rouge turned to education as a means of establishing both legitimacy and political control. However, education under the Khmer Rouge included both *destructive* and *constructive* practices. First, and in conformance with other practices, the Khmer Rouge sought to dismantle the preexisting educational infrastructure. Prior to the Khmer Rouge's rise to power, for example, Cambodia was home to 5,275 primary schools, 146 secondary schools, and 9 institutes of higher education.[68]

Under the direction of the CPK, however, this infrastructure was smashed or demolished. Teachers were "smashed," as anywhere from 75 to 90 percent of all teachers at all levels were killed during the genocide. Most school buildings were destroyed; libraries were emptied and books were burned. Those buildings left standing were often converted to other uses. The Royal University was turned into a farm. Perhaps most symbolic, a former high school (Tuol Svay Prey) was converted into a detention and torture facility—the aforementioned S-21, known as Tuol Sleng prison, at which approximately 14,000 people were detained, tortured, and eventually killed.[69]

Alongside these destructive practices, the Khmer Rouge forwarded a number of (in their view) constructive practices. This marked the second phase of the CPK's educational agenda: the construction of Democratic Kampuchea. Simply stated, the Khmer Rouge leadership proposed a new educational system, one that was intended to promote a national political consciousness and in turn provide legitimacy to Khmer Rouge rule. In fact, the Khmer Rouge explicitly sought to justify their political and economic programs through education.

Education in general, but *geographic* education specifically, is far from a neutral activity. With respect to the latter, it is now well understood that the teaching of geography is important in the development of a political consciousness. Geographic instruction, firstly, provides students with basic knowledge about people and places: the facts and figures of geography, or the traditional "capes-and-bays" form of knowledge that appears on maps and in textbooks. However, there is also a "hidden curriculum" (or subtext) in the teaching of geography. Geographic education may facilitate the construction of "imagined geographies" upon which a social sense of identity can be manufactured and circulated by the state through its institutions, orders, and discourses.[70] This is seen, for example, in the redrawing of political maps following war. And, in fact, following the victory of the CPK, a new map appeared, one that symbolically spoke to the new state of Democratic Kampuchea. The Khmer Rouge's map portrays the administrative divisions of Democratic Kampuchea. At the broadest scale, Democratic Kampuchea was divided into seven geographic zones, identified by cardinal compass directions: North, Northeast, East, Southwest, West, Northwest, and Center. These zones were apparently derived from military designations established by the Khmer Rouge during the war (1970–1975). These zones, significantly, did not conform to any preexisting political division of Cambodia. The Northeast, East, and Southwest zones, for example, included the former eastern portion of Stung Treng Province and the provinces of Ratanakiri, Mondulkiri, Prey Veng, Svay Rieng, eastern Kompong Cham, Kandal, southern Kompong Speu, and Kampot.

The political geography of Democratic Kampuchea as delineated on the map is very significant. Certainly, one sees evidence of the militarized society promoted by the Khmer Rouge. The fact that political divisions, for example, were derived from military necessity is certainly important. However, the map also reveals how the Khmer Rouge sought to erase previous regional identities, to be replaced by an imaginative geography that suppressed regionalism and provincialism in favor of a broader nationalism. The entire political geographic organization of Democratic Kampuchea was based on an abstract system composed of cardinal direction points and numbers and, in the process, the Khmer Rouge's cartography signified "egalitarianism" in that all regions were identical; there was nothing to distinguish one zone from the other.

But how were these new political subjects to be treated within Democratic Kampuchea? How did state governance—as to be taught in school—conform to reality? To provide insight into this question requires an engagement with the practice of medicine under the Khmer Rouge, for as John Harris explains, "Health care is one of the clearest and most visible expressions of a society's attitude to the value of life."[71] In Nazi Germany we see the calculated value of lives not worth living; in Maoist China we see the disregard for all life that was not productive. In Democratic Kampuchea, we witness a combination, whereby the Khmer Rouge made life and death decisions to conform to their own knowledge of lives that were politically loyal, lives that were economically useful, or lives that had no value.

THE PRACTICE OF MEDICINE UNDER THE KHMER ROUGE

On the eve of the genocide, Cambodia exhibited two coexisting medical traditions. On the one hand, traditional forms of medicine were widely practiced while, on the other hand, Western-based "modern" medicinal practices were also available.[72] The distribution of these two systems, however, varied geographically and by population.

The concept of a "medical system" includes both theory and practice. According to Ovesen and Trankell, "theory" in this sense refers to the overall body of medical ideas, or worldview, while "practice" consists of specific techniques and technologies.[73] Both medical theory and practice, moreover, are often—though not always—incorporated within larger political philo-sophical contexts: the ways in which knowledges of medicine and bodies work in concert with knowledges of politics and political bodies. Giorgio Agamben argues that "life and death are not properly scientific concepts but rather political concepts, which as such acquire a political meaning precisely only through a decision."[74]

Cambodia's indigenous (or precolonial) medical tradition was (and remains) based on the *kru khmae* (meaning "Khmer teacher"). Unlike Western medical practice, the indigenous healing process in Cambodia consists of a "diagnosis dialogue." In tandem, the *kru khmae* and the patient coproduce a body of knowledge regarding the latter's health; this knowledge contains physical, social, and cosmological aspects. As such, the traditional practice of medicine in Khmer society is socially, psychologically, and spiritually therapeutic. Of equal importance is that the entire process of diagnosis and therapy must be based on a consensus between the *kru*, the patient, his or her relatives, and the spirits.[75]

The arrival of French colonialism brought with it a Western-based medical practice. Initially, these practices were overseen by the French military and focused on curative (surgical) aspects and some preventative practices. As civilian authorities assumed principal responsibility for the health of its colony, medical practices shifted. Preventive health care, in particular, assumed center stage. French colonial officials were concerned primarily to prevent the spread of infectious diseases such as cholera, smallpox, and the plague. Consequently, substantial investments (both financial and scholarly) were placed on developing "tropical" medicines and specifically vaccinations that would foster the health of French residents in the colony. With limited scope, these preventative practices were extended to the indigenous Cambodians. This was done not so much out of altruism, but rather because it was rational. France's economic interests in Cambodia, certainly, but throughout all of Indochina, hinged on rice production. It was therefore in the financial interests of the colonial authorities to maintain a productive, viable labor force.

Throughout the period of occupation, indigenous medical practices were viewed mostly with indifference by French authorities. In effect, as long as traditional practices did not negatively impact on French policies and practices—such as the vaccination program—they were tolerated. Consequently, the majority of the Khmer were able to maintain their traditional way of life with respect to health. One notable exception was that although the Khmer were generally receptive to French "relief" medicines, as these typically resembled indigenous herbal remedies, surgery was almost universally rejected by the local Khmer.[76]

Following independence in 1953, Western-based medicines were actively promoted through the establishment of medical hospitals, clinics, educational facilities, and pharmaceutical companies. Between 1955 and 1969, for example, the number of hospitals and district clinics rose from 16 to 69; and the number of commune dispensaries increased from 103 to 587. Specifically, in 1959 the French opened *l'Hôpital Calmette* in Phnom Penh—a state-of-the-art research hospital; this was followed with the opening of the Khmer-Soviet Friendship Hospital (*l'Hôpital de l'amitié Khméro-Soviétique*, or "Russian" Hospital) in 1960. *Medical* training also became more widespread. The *l'École des officiers de santé* (a medical school established in 1946) became in 1953 the Royal School of Medicine (*l'École royale de medicine*); this institutional later became the Faculty of Medicine. And, lastly, domestic production of pharmaceuticals began in 1963 with the establishment of the *Pharmacie d'approvision Khmère*.[77]

By 1975 there was a distinct geography to the two medical systems throughout Cambodia. Most Western-based clinics, hospitals, and educational facilities were concentrated in Phnom Penh while the vast majority of Cambodia's population continued to rely on indigenous health practices. In many respects, ironically, this system did not end with the arrival of the Khmer Rouge.

The practice of medicine was not—contrary to some accounts—completely eliminated by the Khmer Rouge. This is a point that cannot be overstated. David Chandler, for example, states that the Khmer Rouge rejected Western-style medicine and refused to import medicine.[78] This, however, is not entirely correct and has a significant bearing on our understanding of political sovereignty. It is more appropriate (as discussed above) to consider the practices forwarded by the Khmer Rouge as a two-phase process of destruction and construction.

The destruction/construction of Democratic Kampuchea's medical system began on or around April 17, 1975. During the initial phase, all hospitals—including doctors, nurses, and patients—were evacuated. Much of the existing medical equipment, as well as medicines, were destroyed. The library of the Medical Faculty in Phnom Penh was burned, as were other medical schools. To this end, Khmer Rouge practices conformed to the broader principles laid out by the CPK. One such practice, as documented by Sokhym Em, was that the revolution was "obliged to abolish all hospitals and their staff left by previous regimes with a view to establishing hospitals of a new style with a socialist character—revolutionary pureness and cleanliness."[79]

Apart from the physical destruction, doctors, nurses, and others who were educated in Western-style medicine were killed. According to some estimates, by 1979 fewer than fifty medical doctors who had practiced medicine before the genocide remained alive. To fill this void, the Khmer Rouge recruited and *trained* its own medical personnel—with predictably tragic results.

The Khmer Rouge had a slogan: "Daughters should grow up to be medical staff, while sons, to be soldiers."[80] Accordingly, young girls—mostly between twelve and fifteen years of age—were recruited from the "base" people living in the provinces. In addition, many other girls actively left home and volunteered to train as nurses or other medical personnel.[81] And finally, many positions within the medical sector were filled by the daughters of high-ranking Khmer Rouge leaders. The four daughters of Ta Mok,[82] for example, became nurses; likewise, all three of Ieng Thirith's children became "social affairs" cadres in Phnom Penh.[83]

Most of those who volunteered or were recruited to work as medical staff were uneducated the Khmer Rouge did not employ any former staff, nor anyone educated, under the prior prerevolutionary system on account of their being part of the "feudal" and "capitalist" class. As Pol Pot explained in a 1978 interview, "The culture of Democratic Kampuchea is a brand-new one . . . containing no reactionary aspects."[84] Just as Cambodia was to be erased so were the constituent elements of the previous medical establishment. A revolution must be total.

Medical training was hurried and rudimentary, lasting at most a matter of weeks. "Nurses" were taught how to give injections and administer pills and other medicines. Since most of the trainees could not read, they were instructed how to "recognize" the symbols (words) that were printed on the bottles.[85] These medical personnel were known *pet padevat* or "revolutionary medics."[86]

Medicines and pharmaceuticals of varying qualities were available throughout Democratic Kampuchea. Despite claims to the contrary, the Khmer Rouge did in fact continue to import supplies from other countries—including those needed for its remade health system. Throughout 1975 and 1976, for example, international trade flourished between Democratic Kampuchea and Thailand. Along with axes, knives, sickles, plow tips, and sugar imported into Democratic Kampuchea, the Khmer Rouge also received sizeable supplies of penicillin, quinine, serum, and vitamins. By late 1976 the CPK also established a trading company, the Ren Fung Company, which conducted trade with Hong Kong. The first shipment—delivered in December 1976—contained nine tons of penicillin, four tons of nivaquin, quinine, and chloroquin, and one ton of vitamins. The Khmer Rouge likewise accepted "gifts" from abroad. Thus, in 1976 the Khmer Rouge received US$12,000 worth of antimalarial drugs sent by the American Friends Service Committee, with Washington's approval, via China.[87]

Imported medicines and medical equipment, dental equipment and dental supplies, and pharmaceuticals were initially stored in warehouses in Phnom Penh and Sihanoukville before being distributed to various hospitals and clinics throughout the country. The most important distribution facility was apparently the K2 warehouse located in Phnom Penh.

The Khmer Rouge also established a series of factories to "produce" their own medicines and pharmaceuticals. At least four pharmaceutical factories are known to have been in operation in Phnom Penh, designated as P1, P3, P4, and P6. The factory known as P1 produced traditional medicines while the other three produced penicillin, serum, setropharine, and vitamins B1, B6, and B12.[88] The efficacy of these medicines and pharmaceuticals was, not surprisingly, of dubious quality. One type of medicine produced was based on indigenous practices and local *kru* were recruited to oversee production. These medicines were used to "cure" fever, headache, stomachaches, and fainting. Chinese advisors often provided guidance in the manufacture of indigenous medicines. Raw materials included plant roots, tree bark, sap, and other "natural" compounds. Young girls would mix these materials together and shape them into small pills, which became widely known as *rabbit dropping* medicine because of their appearance—and effectiveness.[89]

Chinese advisors likewise oversaw the production of Western-based medicines and pharmaceuticals; these were produced with the raw materials imported from Thailand, Hong Kong, and elsewhere. The quality of these drugs varied widely; their distribution was selective. As with medicinal care overall, drugs were rationed throughout Democratic Kampuchea, with the Khmer Rouge calculating who was to receive which medicines.

Figure 4.10.   Young girls preparing "rabbit dropping" medicine. *Source:* Documentation Center of Cambodia

Figure 4.11.   Young girls preparing "rabbit dropping" medicine. *Source:* **Documentation Center of Cambodia**

A hierarchical system of hospitals, clinics, and other health facilities was established throughout Democratic Kampuchea. At this point, unlike the detailed understandings of the Nazi medical program, the precise coordinates and arrangement of the Khmer system remain poorly understood. It is known that the two major hospitals in Phnom Penh—although initially evacuated and ransacked—were later put into operation: the Russian Hospital (renamed the 17 April Hospital) and the Calmette Hospital. Outside the capital city, provincial hospitals existed; these generally had a few fully trained doctors, some Chinese advisors, and "decent" medicinal supplies. Conditions deteriorated as one proceeded down the hierarchy. At the district level, health clinics were rarely staffed by trained medicinal practitioners and medicines consisted mostly of domestically produced pharmaceuticals—the "rabbit dropping" medicine. The lowest level consisted of *munti pet* (infirmaries or small clinics). In typical fashion, the Khmer Rouge simply appropriated suitable buildings, such as *wats* and other religious sites, to serve as infirmaries.[90] These would be staffed by *pet padevat* and were poorly supplied at best.

The Khmer Rouge did permit the practice of indigenous healers—the *kru khmae*. However, the ability of the *kru* to conduct their services was severely limited. While able to distribute traditional medicines, the Khmer Rouge forbade the *kru khmae* from performing the spiritual component of health care; likewise, *kru* were not allowed to use mantras or offerings, and the family's role in health care was prohibited.[91]

Both traditional and Western-based medicinal practices were thus available within Democratic Kampuchea. The crucial point, as Ovesen and Trankell correctly note, is that proper hospitals, adequately trained doctors and nurses, and effective pharmaceuticals were not for everyone.[92] As discussed later, it was this differential among Democratic Kampuchea's population that became in large part the difference between life and death.

During its reign, the Khmer Rouge constructed—and normalized—many different populations that were subjected to differential treatment: a calculated management of life through the provision of health care. Some populations were ethnically defined. The Khmer Rouge, for example, targeted all persons who were perceived as racially distinct, including the Viet-Khmer, the Sino-Khmer, and the Muslim Cham. Other marginalized populations were based on previous occupation (e.g., soldiers and officials of the previous government). Most prevalent, however, was the political distinction between "new" and "old" people. "Old" people (also known as "base" people) were those who either participated in the revolution and/or lived in Khmer Rouge–controlled zones during the war. According to Khamboly Dy, the Khmer Rouge classified base people as "full-rights" people, meaning that these people were allowed to vote[93] and run for elections; they could become chiefs of cooperatives; and they generally had access to better food rations and, significantly, medical care. Conversely, "new" people—also designated as "17 April" people—consisted of those people who were evacuated from the cities and towns following the Khmer Rouge victory. New people received less food, were treated more harshly, had fewer rights, and were directly killed more readily than old people.[94] Much as the Jews, homosexuals, and Roma of Nazi Germany, the "new" people embodied the concept of *homo sacer*—a life that could be killed with impunity and whose death held no significance.

Both the Calmette and Russian Hospitals were reserved primarily for the highest Khmer Rouge leadership—the *Angkar loeu*.[95] In the Russian Hospital, for example, CPK leaders would stay in private rooms with attached bathrooms; they would be assisted by personal servants and received choice (and adequate) food; and they were treated with imported medicines administered by Chinese doctors.[96] And while conditions were decidedly worse at the provincial and district level facilities, none compared with the malpractice of medicine found at the *munti pet*. Here, the *pet padevat* both learned and practiced health care—often with tragic results.

Young girls, barely into their teens, were largely responsible for diagnosis and treatment at the *munti pet*. Beyond certain obvious symptoms of ailment or illness—fevers, body aches, diarrhea, and open wounds, these medics could do nothing. They administered rabbit dropping medicines (when available) and gave injections of any liquids available—coconut milk, chicken soup. Predictably, the administration of improper medicines caused many patients to suffer adverse reactions or die.[97] But this too was consistent with CPK policy. As Pol Pot explained in a 1978 interview, "We have to establish a research team to do research and conduct experiments on traditional drugs. Even though we do not have proper formulas, we can still produce them. We are practicing self-reliance in medicine."[98]

The overall effect of CPK policies and practices was devastating. Between 1975 and 1979 approximately two million people died in Cambodia—approximately one-quarter of the country's pre-genocide population. Debate remains as to the proportion of deaths—of those who died from direct violence (i.e., execution) and those who died from indirect means (e.g., starvation, disease, exhaustion, and exposure). A consensus is developing, however, that deaths were roughly proportional, that is, approximately 50 percent of people died due to direct violence while another 50 percent died from other causes.[99]

Figure 4.12.   Mass grave site at one of Cambodia's many killing fields. *Source:* Documentation Center of Cambodia

"LETTING DIE" IN THE CAMBODIAN GENOCIDE

It is well established that the Khmer Rouge enacted specific policies of torture and execution that led to significant loss of life. My immediate concern, however, focuses on those people who *were not killed directly* by the Khmer Rouge—but who died nonetheless because of specific policies and practices enacted as a form of statecraft. How do we account for the millions of people who died from starvation, exhaustion, or exposure to the elements?

This question relates directly to a long-standing moral problem for both philosophers and medical ethicists: Is there a moral distinction between killing and letting die? Can we argue that the leaders of the Khmer Rouge are in fact morally culpable for these deaths; alternatively, are we left with the uncomfortable conclusion that the leadership of the Khmer Rouge simply let die masses of people?—a claim, much like those in Maoist China, that some of the perpetrators maintain.

At the outset of this book I suggested that an engagement with genocide may expand our understanding of the politics of sovereignty and the calculated management of life and death. Specifically, my focus is on the moral geographies that encompass the sovereign's right to kill, to foster life, or to disallow life to the point of death. To do so requires that we pry open, much as we did in our discussion of China, the philosophical and ethical question of life and death from the standpoint of state sovereignty. Thus, to paraphrase the above question: Is it worse for the state to kill someone than to allow them to die?

Intuitively, we may presume that the act of killing is morally worse than letting die. Such a presumption hinges on our understanding of agency: to kill is an action, whereas letting die is an omission, or lack of action. Furthermore, this moral distinction is founded upon a distinction between "negative" and "positive" duties. As Cartwright identifies, we have duties not to harm others, which require restraint—these are termed negative duties.[100] We also have positive duties, whereupon we have duties to help others. According to Cartwright, since negative duties not to injure carry more weight than positive duties to aid, infringements of the former, like killing, are often considered more grave than infringements of the latter, such as letting die.[101] From this perspective, state practices that *intentionally* injure or kill are thus significantly more serious than the state's failure to enact positive practices that aid its citizens. And indeed, it is this distinction that designates extrajudicial murder or war-rape as "crimes against humanity," while the lack of providing adequate health care is not considered to be a similar crime. In the former situation, the state may be found guilty through its perpetration of specific practices that directly kill people; in the latter situation, however, the lack of action is considered to be neither morally nor legally wrong.

Counter to this argument, I suggest that *there is no moral distinction between killing and letting die*. I concur with Rachels who finds that the "bare fact that one act is an act of killing, while another act is an act of 'merely' letting someone die, is not a morally good reason in support of the judgment that the former is worse than the latter."[102] Consequently, state intervention and *nonintervention* must be accounted for when discussing political sovereignty and the calculated management of life and death. I premise my argument on two elements: (1) the conditions of letting die and (2) the concept of "to refrain."

According to Cartwright, a plausible account of the difference between killing and letting die may be that one kills someone if one initiates a causal sequence that ends in one's death, whereas one lets another die if one allows an already existing casual sequence to culminate in that person's death.[103] In effect, Cartwright is addressing the broader context, or conditions, that may result in death. Geographers have addressed these conditions at length, for they fall under the rubric of structural violence. From this perspective, however, states (should) have a moral and legal responsibility for the creation of conditions that lead to starvation, malnutrition, or exposure to pollutants. Within the context of the Cambodian genocide, therefore, we may argue that those social and environmental conditions that led to famine, disease, exhaustion, and exposure to the elements were the direct result of specific practices undertaken by the Khmer Rouge. The collectivization of agriculture, the extreme rationing of foods, the mandatory dawn-to-dusk work hours: these were the conditions that made death possible. Those with power in Democratic Kampuchea did not simply let die. Rather, the leadership created those conditions—the labor camps for example—that led directly to the deaths of millions of people.

An argument predicated on conditionality—that of structural violence—is necessary but insufficient. A defendant of the Khmer Rouge *might* argue that the deaths resultant from these conditions were unfortunate but unintended. Instead, those practices—the collectivization of agriculture, the establishment of labor camps—were enacted *to foster life*. Consider, for example, the CPK's Four-Year Plan for 1977–1980, developed between July 21 and August 2, 1976. Although never published, and apparently not widely implemented, the document provides insight into key policies and practices envisioned by the Standing Committee of the CPK.[104] Thus, a stated objective of the CPK was to "produce rice for food to raise the standard of living of the people" and "to obtain capital" for the purchase of imports.[105] However, according to the document, the CPK recognized several immediate limitations, such as the lack of existing infrastructure to facilitate increased rice production. These *structural* limitations were used to justify the establishment of labor camps—to build needed canals, reservoirs, and irrigation pumps—and to initiate food quotas.

A defendant of the Khmer Rouge might also argue that the establishment of a medical system likewise supports the contention that CPK practices were designed to foster life. I disagree, noting that the practice of medicine does at one level suggest an attempt to foster life, but it also demonstrates that the Khmer Rouge were willing to disallow life to the point of death. That the state at this point must be held morally (and legally) accountable for these deaths. It is here that the second component, that of the concept "to refrain" enters the picture.

Recall the presumption that killing is worse than letting die. According to Rachels, we often think this, not because we overestimate how bad it is to kill, but because we underestimate how bad it is to let die.[106] It is important to acknowledge that letting die does not necessarily equate with a lack of agency. Indeed, *we let someone die.* In other words, we refrain from preventing someone from dying. This, I believe, strikes at the heart of both Foucault's and Agamben's concern with respect to the valuation of life and the disallowing of life to the point of death. How does this operate? Let us consider more closely Green's discussion on the concept "to refrain."

Following Green, "letting die is not simply not doing something to prevent death."[107] Rather, to let die is to fail to act—it is intentional, it is to refrain from acting otherwise. This argument, however, rests on three conditions: ability, opportunity, and awareness. First, consider ability. According to Green, to refrain from performing an action involves not performing that action *but having the ability to perform it.*[108] Stated in the form of a question: Is an individual in a position to prevent a death but, through his act or inaction, fails to do so? In Democratic Kampuchea, the CPK leadership had the ability to save lives—to prevent deaths. This is evidenced by the continuance of hospitals, by the presence of indigenous medicines and the importation of Western-based medicines, and by the presence of skilled doctors. However, the CPK leadership refrained from providing *equal* access to these services. In a stark calculated management of life, the Khmer Rouge determined who was allowed to live and whose lives were disallowed to the point of death. The hierarchical construction of populations by the CPK established the conditions whereby top-ranking Khmer Rouge leaders (often) had access to life-savings techniques and practices; rank-and-file members ("base" people) had less access; and those classified as "new" people, or "17 April" people, had access, at best, to inferior medicines and medical treatment. And in an ironic twist, the sick and the injured were often viewed as *least deserving* of medical care. Henri Locard discussed how the Khmer Rouge viewed the "sick" and the "disabled" as impediments to the revolution, to the state. According to

Locard, "the sick could not be anything other than malingerers, because they could or would not work, and therefore, were sabotaging the revolution."[109] The following slogans capture these sentiments: "We absolutely must remove [from society] the lazy; it is useless to keep them, else they will cause trouble. We have to send them to hell"; "The sick are victims of their own imagination"; and "We must wipe out those who imagine they are ill, and expel them from society!" Locard explains that all Cambodians "heard these rebukes shouted at them with violence and unusual harshness every time they had a fever or fell ill."[110] In short, for the vast majority of the people of Democratic Kampuchea, by denying access to available medicines and medical techniques, the CPK leadership disallowed life to the point of death.

A second condition rests on the idea of opportunity: Is it logically possible that a person has the opportunity to prevent death? Take an obvious geographic example. A doctor living in Los Angeles may have the ability—the skills—to save a life of a person suffering a heart attack in New York—but not, because of the spatial separation, have an opportunity to do so. Within Democratic Kampuchea, as the work conducted by Sokhym Em, Laura Vilim, Jan Ovesen, and Ing-Britt Trankell among others has documented, the CPK leaders had the opportunity to provide better medical care. Furthermore, these leaders had the opportunity to rethink and thus improve the deplorable working and living conditions discussed above.

Awareness constitutes the final condition: Did the CPK leadership know of the conditions? Were they aware that their policies and practices were resulting in widespread morbidity and mortality? Empirical evidence, not surprisingly, is hard to come by. Certainly those Khmer Rouge leaders deny any culpability for the death of millions of Cambodians. However, circumstantial evidence at the Documentation Center of Cambodia (DC-CAM) points to the activities of Ieng Thirith. As minister of social affairs and action, and head of Democratic Kampuchea's Red Cross, she was responsible for implementing the medical policies and food rationing of the country. Evidence suggests that she in fact investigated health issues throughout the country and was thus aware of the deplorable conditions that led to so many thousands of deaths. Furthermore, addition evidence is mounting that other top-ranking members of *Angkar loeu* were well aware of the grievous effects that their policies and practices were having on the Cambodian populace.

To sum: A person can be said to refrain from performing an action if, and only if, he or she has the ability and opportunity to perform the action—and is aware of this—but does not perform the act—and is also aware of this.[111] Consequently, we may conclude that while killing may be unintentional (e.g., manslaughter), letting die cannot be. The Khmer Rouge leadership created the conditions—through their policies and practices—that led to widespread famine, malnutrition, exhaustion, and exposure. These leaders, moreover, were aware of the deleterious effects of their policies and practices; they had the opportunity and ability to introduce corrective measures, but did not act. We are left with the conclusion that one who refrains from preventing death does not "simply" let die, but in fact *kills*.[112]

CONCLUSIONS

Through brutal systems of terror, the Khmer Rouge attempted to fix the spatiality of Democratic Kampuchea to conform to their own geographical imagination.[113] Any resistance, any self-identity, any contestation between the subject and the state was to be eliminated. This was accomplished through both direct killing and letting die: torture, execution, exhaustion, starvation. Those who in the Khmer Rouge's calculations of lives worth living were deemed extraneous were subject to extreme practices of material deprivation and extermination.

Karl Jackson refers to the Cambodian revolution as a "revolution by eradication."[114] Conceptually, we understand the eradication of people as "ethnic cleansing." Constituting their geographical imagination, the Khmer Rouge sought to construct a new, pure utopian society not through transforming past geographies, but by erasing those spaces and starting anew. The mass violence unleashed by the Khmer Rouge was not, from their perspective, irrational or insane: it was extraordinarily ordinary. The violence meted out by the Khmer Rouge was deliberate and, in some respects, very methodical. From the perspective of the Khmer Rouge, genocide was functional.

Within the geographical imagination of the Khmer Rouge, the killing and allowance of death had a clear and distinct purpose: a systematic eradication of bodies that did not conform to the imagined geographies of Pol Pot and his followers. And on this level, the Cambodian genocide resonates with the Nazi-led Holocaust. Whether people died from starvation, lack of medical care, or execution was inconsequential. Haing Ngor, a survivor of the Cambodian genocide, explains that these people, to the Khmer Rouge, "weren't quite people. [They] were lower forms of life, because [they] were enemies. Killing [them] was like swatting flies, a way to get rid of undesirables."[115]

The construction of a pure society—written onto pages erased clean from the stains of previous regimes—requires the eradication and elimination of those individuals calculated as nonconformist and thus not worthy of inclusion. These bodies, conceived as essentially separate, could not (or only rarely), in the geographical imagination of the Khmer Rouge, be reeducated, rehabilitated, and reintroduced into society. There emerged a perceived necessity to both kill and let die the unwanted, the impure. To be sure, these bodies would be worked—to death if necessary. Ultimately, as François Ponchaud explains, the purge of the Cambodian population "was, above all, the translation into action of a particular vision of man [*sic*]: a person who has been spoiled by a corrupt regime cannot be reformed, he must be physically eliminated from the brotherhood of the pure."[116]

Life had little value for the Khmer Rouge leadership—beyond the purpose of serving the state. This is expressed brutally in the Khmer Rouge slogan: "To die is banal for the one who fights heroically."[117] In Democratic Kampuchea, the individual body was inconsequential. As a collective, as a "population," bodies had worth only insofar as they put into action the geographical imagination of a soon-to-be achieved utopia. Populations were to be productive, politically loyal—or nothing. This is why Pol Pot, on his deathbed, continued to exhibit no remorse for the death he unleashed on Cambodia. This is why so many Nazi officials refused to express remorse for their actions. And this is why Mao and his cadre admitted no wrongdoing. In all instances, the "state" did not kill *anyone*; the state simply removed the unworthy that posed a threat. State-sanctioned violence, both direct and structural, was considered just.

NOTES

1. Sarom Prak, "The Unfortunate Cambodia," in *Children of Cambodia's Killing Fields: Memoirs by Survivors*, compiled by Dith Pran and edited by Kim DePaul (New Haven, CT: Yale University Press, 1997), 67–71.

2. Prak, "The Unfortunate Cambodia," 71.

3. David P. Chandler, *A History of Cambodia*, 3rd ed. (Boulder, CO: Westview Press, 2000).

4. David P. Chandler, *The Tragedy of Cambodian History: Politics, War, and Revolution Since 1945* (New Haven, CT: Yale University Press, 1991), 108.

5. Gareth Porter, "Vietnamese Communist Policy toward Kampuchea, 1930–1970," in *Revolution and Its Aftermath: Eight Essays*, edited by David P. Chandler and Ben Kiernan (New Haven, CT: Yale University Southeast Asian Studies, 1983), 57–98; at 60.

6. Ben Kiernan, "The Cambodian Genocide 1975–1979," in *Century of Genocide: Critical Essays and Eyewitness Accounts*, 3rd ed., edited by Samuel Totten and W. S. Parsons (New York: Routledge, 2009), 341–75; at 344.

7. Ben Kiernan, *How Pol Pot Came to Power: A History of Communism in Kampuchea, 1930–1975* (London: Verso, 1985); Kenneth M. Quinn, "The Pattern and Scope of Violence," in *Cambodia 1975–1978: Rendezvous with Death,* edited by Karl D. Jackson (Princeton: Princeton University Press, 1989), 179–208; and Ben Kiernan, *The Pol Pot Regime: Policies, Race and Genocide in Cambodia under the Khmer Rouge, 1975–1979* (New Haven, CT: Yale University Press, 1996).

8. Thomas Clayton, "Building the New Cambodia: Educational Destruction and Construction Under the Khmer Rouge, 1975–1979," *History of Education Quarterly* 38(198): 1–16; David M. Ayers, *Anatomy of a Crisis: Education, Development, and the State in Cambodia, 1953–1998* (Chiang Mai, Thailand: Silkworm Press, 2000); Thomas Clayton, "Re-Orientations in Moral Education in Cambodia Since 1975," *Journal of Moral Education* 34(2005): 505–17; James A. Tyner, "Imagining Genocide: Anti-Geographies and the Erasure of Space in Democratic Kampuchea," *Space and Polity* 13(2009): 9–20; and James A. Tyner, "Genocide as Reconstruction: The Political Geography of Democratic Kampuchea," in *Reconstructing Conflict: Integrating War and Post-War Geographies*, edited by Scott Kirsch and Colin Flint (Aldershot, UK: Ashgate, 2011), 49–66.

9. Quoted in various sources.

10. Henri Lefebvre, *The Production of Space* (Oxford: Blackwell, 1991), 54.

11. Derwent Whittlesey, "Sequent Occupance," *Annals of the Association of American Geographers* 19(1929); 162–65.

12. Lefebvre, *Production of Space*, 46–47.

13. Lefebvre, *Production of Space*, 47.

14. David P. Chandler, *Brother Number One: A Political Biography of Pol Pot*, rev. ed. (Chiang Mai, Thailand: Silkworm Press, 2000), 209; see also Serge Thion, "The Cambodian Idea of Revolution," in *Revolution and Its Aftermath in Kampuchea*, edited by Chandler and Kiernan, 10–33; and Philip Short, *Pol Pot: Anatomy of a Nightmare* (New York: Henry Holt & Co., 2004).

15. Thion, "Cambodian Idea," 25.

16. Thion, "Cambodian Idea."

17. Karl D. Jackson, "The Ideology of Total Revolution," in *Cambodia 1975–1978: Rendezvous with Death*, edited by Karl D. Jackson (Princeton, NJ: Princeton University Press, 1989), 59. It is a common misreading of the Khmer Rouge that the Party attempted to return Cambodia to the glories of the Angkorian empire (ninth through fourteenth centuries). Ian Brown, for example, in an otherwise highly informative book, writes that the "clock was to be turned back to an age without money, organized education, religion, and books." This is patently *wrong*. The Khmer Rouge in fact produced their own currency, developed educational curricula, established state ideology as religion, and published textbooks to be used in schools. See Ian Brown, *Cambodia* (Herndon, VA: Stylus Publishing, 2000), 24.

18. Kenneth M. Quinn, "The Pattern and Scope of Violence," in *Cambodia, 1975–1978: Rendezvous with Death*, edited by Karl D. Jackson (Princeton: Princeton University Press, 1989), 179–208; at 181.

19. Kiernan, *How Pol Pot Came to Power*, 371.

20. Kiernan, *How Pol Pot Came to Power*, 384.

21. François Pouchad, "Social Change in the Vortex of Revolution," in *Cambodia, 1975–1978: Rendezvous with Death*, edited by Karl D. Jackson (Princeton: Princeton University Press, 1989), 151–77; 161.

22. See James A. Tyner, *Killing of Cambodia: Geography, Genocide and the Unmaking of Space* (Aldershot, UK: Ashgate, 2008), 150–68.

23. Henri Locard, *Pol Pot's Little Red Book: The Sayings of Angkar* (Chiang Mai, Thailand: Silkworm Press, 2004), 143.

24. Locard, *Pol Pot's Little Red Book*, 144.

25. Quinn, "The Pattern and Scope of Violence," 193.

26. Prak, "The Unfortunate Cambodia," 68.

27. Kalyanee Mam, "The Endurance of the Cambodian Family under the Khmer Rouge Regime: An Oral History," in *Genocide in Cambodia and Rwanda: New Perspectives*, edited by Susan E. Cook (New Haven, CT: Yale Center for International and Area Studies, Genocide Studies Program Monograph Series, no. 1, 2004), 127–71; at 146.

28. Mam, "The Endurance of the Cambodian Family," 146.

29. Peg LeVine, *Love and Dread in Cambodia: Weddings, Births, and Ritual Harm under the Khmer Rouge* (Singapore: National University of Singapore Press, 2010), 29.

30. Prak, "The Unfortunate Cambodia," 68.

31. Michel Foucault, *"Society Must Be Defended": Lectures at the Collège de France, 1975–76*, translated by David Macey (New York: Picador, 2003), 256.

32. Michel Foucault, *Discipline and Punish: The Birth of the Prison*, translated by A. Sheridan Smith (New York: Vintage Books, 1979), 215.

33. "Excerpted Report on the Leading Views of the Comrade Representing the Party Organization at a Zone Assembly," translated by Ben Kiernan, in *Pol Pot Plans the Future: Confidential Leadership Documents from Democratic Kampuchea, 1976–1977*, edited by David P. Chandler, Ben Kiernan, and Chanthou Boua (New Haven, CT: Yale University Southeast Asia Studies, 1988), 9–35.

34. "Excerpted Report," 20.

35. "Excerpted Report," 20.

36. Chandler, *History of Cambodia*, 211.

37. "Excerpted Report," 20–21.

38. "The Party's Four-Year Plan to Build Socialism in All Fields, 1977–1980," introduced by David Chandler and translated by Chanthou Boua, in *Pol Pot Plans the Future: Confidential Leadership Documents from Democratic Kampuchea, 1976–1977*, edited by David P. Chandler, Ben Kiernan, and Chanthou Boua (New Haven, CT: Yale University Southeast Asian Studies, 1988), 3–119.

39. "Four-Year Plan," 110–11.

40. May Ebihara, "Revolution and Reformulation in Kampuchean Village Culture," in *The Cambodian Agony*, 2nd ed., edited by David A. Ablin and Marlowe Hood (New York: M.E. Sharpe, 1990), 16–61. See also Mam, "The Endurance of the Cambodian Family," 129.

41. Ebihara, "Revolution and Reformulation," 18.

42. Ebihara, "Revolution and Reformulation," 26.

43. Teeda Butt Mam, "Worms from Our Skin," in *Children of Cambodia's Killing Fields: Memoirs by Survivors*, compiled by Dith Pran and edited by Kim DePaul (New Haven, CT: Yale University Press, 1997), 11–17; at 14.

44. Foucault, *Discipline and Punish*, 143.

45. "Four-Year Plan," 112.

46. "Four-Year Plan," 112.

47. Quinn, "Pattern and Scope of Violence," 182.

48. Roeun Sam, "Living in Darkness," in *Children of Cambodia's Killing Fields: Memoirs by Survivors*, compiled by Dith Pran and edited by Kim DePaul (New Haven, CT: Yale University Press, 1997), 73–81.

49. Sam, "Living in Darkness," 74.

50. Mam, "The Endurance of the Cambodian Family," 134.

51. Mam, "The Endurance of the Cambodian Family," 134–35.

52. Ebihara, "Revolution and Reformulation," 60.

53. Mam, "The Endurance of the Cambodian Family," 143.

54. Locard, *Pol Pot's Little Red Book*, 91.

55. Locard, *Pol Pot's Little Red Book*, 92–93.

56. Locard, *Pol Pot's Little Red Book*, 91.

57. Haing Ngor (with R. Warner), *Survival in the Killing Fields* (New York: Carroll & Graf Publishers, 1987), 277.

58. Ouk Villa, "A Bitter Life," in *Children of Cambodia's Killing Fields: Memoirs by Survivors*, compiled by Dith Pran and edited by Kim DePaul (New Haven, CT: Yale University Press, 1997), 115–21; at 116.

59. Sam, "Living in Darkness," 74–75.

60. Sam, "Living in Darkness," 75.

61. Khamboly Dy, *A History of Democratic Kampuchea (1975–1979)* (Phnom Penh: Documentation Center of Cambodia, 2007).

62. See Ngor, *Survival in the Killing Fields*, 266.

63. David P. Chandler, *Voices from S-21: Terror and History in Pol Pot's Secret Prison* (Berkeley: University of California Press, 1999).

64. Peter J. Taylor, "God Invented War to Teach Americans Geography," *Political Geography* 23(2004): 487–92.

65. See, for example, Virginie Mamadouh, "Geography of War, Geographers and Peace," in *Geography of War and Peace: From Death Camps to Diplomats*, edited by Colin Flint (Oxford: Oxford University Press, 2005), 26–60.

66. Susan Schulten, *The Geographical Imagination in America, 1880–1950* (Chicago: University of Chicago Press, 2001), 1–2.

67. Schulten, *Geographical Imagination*, 2–3.

68. Clayton, "Building the New Cambodia," 5.

69. See, for example, Chandler, *Voices from S-21*.

70. J. Huckle, "Geography and Schooling," in *The Future of Geography*, edited by Ron Johnston (London: Methuen, 1985), 291–306; F. Kashani-Sabet, "Picturing the Homeland: Geography and National Identity in Late Nineteenth- and Early Twentieth-Century Iran," *Journal of Historical Geography* 24(1998): 413–30; and J. Morgan, "Imagined Country: National Environmental Ideologies in School Geography Textbooks," *Antipode* 35(2003): 444–62.

71. John Harris, *The Value of Life: An Introduction to Medical Ethics* (New York: Routledge, 1985), 1.

72. Jan Ovesen and Ing-Britt Trankell, *Cambodians and Their Doctors: A Medical Anthropology of Colonial and Post-Colonial Cambodia* (Copenhagen: Nordic Institute of Asian Studies, 2010).

73. Ovesen and Trankell, *Cambodians and Their Doctors*, 5.

74. Giorgio Agamben, *Homo Sacer: Sovereign Power and Bare Life* (Stanford: Stanford University Press, 1998), 164.

75. Ovesen and Trankell, *Cambodians and Their Doctors*, 9.

76. Ovesen and Trankell, *Cambodians and Their Doctors*, 69.

77. Ovesen and Trankell, *Cambodians and Their Doctors*, 79–81.

78. Chandler, *Tragedy of Cambodian History*, 249 and 259.

79. Sokhym Em, "Female Patients," *Searching for the Truth* 33(2002): 25–29; at 26.

80. Sokhym Em, "Revolutionary Female Medical Staff in Tram Kak District, Part I" *Searching for the Truth* 34(2002): 24–27; at 25.

81. According to Sokhym Em the motivation to volunteer varied. For some girls, it was a means of avoiding oppression by local authorities; for others it was to avoid fighting on the battlefield. See Em, "Revolutionary Female Medical Staff," 25.

82. Ta Mok (meaning "Grandfather" Mok) was a military general and member of the Standing Committee of the CPK's Central Committee. His real name was Chhit Choeun.

83. Em, "Revolutionary Female Medical Staff," 25.

84. Em, "Revolutionary Female Medical Staff," 25.

85. Sokhym Em, "Revolutionary Female Medical Staff in Tram Kak District, Part II" *Searching for the Truth* 35(2002): 17–19.

86. Ovesen and Trankell, *Cambodians and Their Doctors*, 91.

87. The Khmer Rouge also conducted trade with Yugoslavia, Albania, Romania, China, and North Korea. See Kiernan, *The Pol Pot Regime,* 145–46; and Ovesen and Trankell, *Cambodians and Their Doctors*, 209.

88. Ovesen and Trankell, *Cambodians and Their Doctors*, 107.

89. Sokhym Em, "Rabbit Dropping Medicine," *Searching for the Truth* 30(2002), 22–23; at 22.

90. Ovesen and Trankell, *Cambodians and Their Doctors*, 91.

91. Ovesen and Trankell, *Cambodians and Their Doctors*, 92.

92. Ovesen and Trankell, *Cambodians and Their Doctors*, 87.

93. Only one election was held in the four-year history of Democratic Kampuchea—on March 20, 1976. According to Dy, this election (not surprisingly) was not in accordance with international standards, and those voted in as members of the National Assembly were not announced publicly. See Khamboly Dy, *A History of Democratic Kampuchea (1975–1979)* (Phnom Penh: Documentation Center of Cambodia, 2007).

94. Alexander L. Hinton, *Why Did They Kill? Cambodia in the Shadow of Genocide* (Berkeley: University of California Press, 2005), 86.

95. The Russian Hospital was reserved for adults while the Calmette Hospital was a children's hospital. Both served almost exclusively members of the CPK and other high-ranking political and military leaders.

96. Ovesen and Trankell, *Cambodians and Their Doctors*, 102.

97. Em, "Female Patients," 26.

98. Laura Vilim, "'Keeping Them Alive, One Gets Nothing; Killing Them, One Loses Nothing': Prosecuting Khmer Rouge Medical Practices as Crimes against Humanity," http://www.dccam.org/Tribunal/Analysis/pdf/Prosecuting_Khmer_Rouge_Medical_Practices_as_Crimes_Against_Humanity.pdf (accessed December 2010).

99. See, for example, Michael Vickery, "How Many Died in Pol Pot's Kampuchea?" *Bulletin of Concerned Asian Scholars* 20(1988): 377–85; Ben Kiernan, "The Genocide in Cambodia, 1979–1979," *Bulletin of Concerned Asian Scholars* 22(1990): 35–40; Patrick Heuveline, "'Between One and Three Million': Towards the Demographic Reconstruction of a Decade of Cambodian History," *Population Studies* 52(1998): 49–65; Ben Kiernan, "The Demography of Genocide in Southeast Asia: The Death Tolls in Cambodia, 1975–1979, and East Timor, 1975–1980," *Critical Asian Studies* 35(2003): 585–97; Damien de Walque, "Selective Mortality during the Khmer Rouge Period in Cambodia," *Population and Development* 31(2005): 351–68; Craig Etcheson, *After the Killing Fields: Lessons from the Cambodian Genocide* (Lubbock: Texas Tech University Press, 2005); and Damien de Walque, "The Socio-Demographic Legacy of the Khmer Rouge Period in Cambodia," *Population Studies* 60(2006): 223–31.

100. W. Cartwright, "Killing and Letting Die: A Defensible Distinction," *British Medical Bulletin* 52(1996): 354–61; at 358.

101. Cartwright, "Killing and Letting Die," 360.

102. J. Rachels, "Killing and Starving to Death," *Philosophy* 54(1979): 159–71; at 164.

103. Cartwright, "Killing and Letting Die," 354.

104. David P. Chandler, "Introduction: 'The Party's Four-Year Plan to Build Socialism in All Fields, 1977–1980,'" in *Pol Pot Plans the Future: Confidential Leadership Documents from Democratic Kampuchea, 1976–1977*, edited by David P. Chandler, Ben Kiernan, and Chanthou Boua (New Haven, CT: Yale University Southeast Asia Studies, 1988), 36–119; at 36.

105. Chandler, "Introduction," 51.

106. Rachels, "Killing and Starving," 160.

107. O. H. Green, "Killing and Letting Die," *American Philosophical Quarterly* 17(1980): 195–204; at 196.

108. Green, "Killing and Letting Die," 196.

109. Locard, *Pol Pot's Little Red Book*, 187–188.

110. Locard, *Pol Pot's Little Red Book*, 188. Locard (p. 188) explains that the reference to "imagination" is somewhat ambiguous. In Khmer Rouge parlance, the term could mean "ideological frame of mind." Consequently, those accused of malingering were behaving so because their minds were infected with the ideology of the old society; they were not pure of mind or heart. This is made clear in the slogan: "If you have the disease of the old society, take a dose of Lenin as medication."

111. Green, "Killing and Letting Die," 197.

112. Green, "Killing and Letting Die," 198.

113. See, for example, Tyner, *The Killing of Cambodia,* 160.

114. Karl D. Jackson, "The Ideology of Total Revolution," in *Cambodia, 1975–1978: Rendezvous with Death*, edited by Karl D. Jackson (Princeton: Princeton University Press, 1989), 37–78; at 56.

115. Ngor, *Survival in the Killing Fields*, 247.

116. Quoted in Quinn, "The Pattern and Scope of Violence," 185.

117. Locard, *Pol Pot's Little Red Book*, 210.

Chapter Five

Everyday Death and the State

In *Geography and the Geographical Imagination* I began with a simple question: What might genocide say about the spatiality of life and death? My concern, broadly stated, follows Joe Painter's notion of the "prosaic manifestations of state processes" and, specifically, the "ways in which everyday life is permeated by stateness in various guises."[1]

Ben Highmore maintains that for many theorists, the "everyday" includes "those most repeated actions, those most traveled journeys, those most inhabited spaces that make up, literally, the day to day."[2] Philip Wander concurs, writing that "everyday life" "refers to the dull routine, the ongoing go-to-work, pay-the-bills, homeward trudge of daily existence. It indicates a sense of being in the world beyond philosophy, virtually beyond the capacity of language to describe, that we know simply as the grey reality of enveloping all we do."[3] Our everyday is "ordinary"; that endless string of days that we scarcely remember. We may, for example, recall with great clarity that day, many years ago, when Grandmother died. But the day before her death? Two days before? No doubt we ate, perhaps went to work—but beyond that? Much of our life, it seems, is a collection of unremarkable days filled with seemingly repetitive acts—not events.

According to Josh Inwood a key point to be drawn is that "experience" is central to an understanding of the everyday.[4] David Delaney, for example, forwards the idea of *geographies of experience*. He explains that "our lives are, in a sense, made of time. But we are also physical, corporeal, mobile beings. We inhabit a material, spatial world. We move through it. We change it. It changes us. Each of us is weaving a singular path through the world. The paths that we make, the conditions under which we make them, and the experiences that those paths open up or close off are part of what makes us who we are."[5] In short, who we are is not entirely of our own choosing; rather,

153

we are constituted through our everyday experiences. As Wander concludes, the "shape and content of our lives is the product of a number of decisions in which we do not participate and about which we may or may not be aware."[6]

Henri Lefebvre forwards a more critical understanding of the "everyday." Lefebvre writes that "people are born, live and die. They live well or ill; but they live in everyday life, where they make or fail to make a living either in the wider sense of surviving or not surviving, or just surviving or living their lives to the full. It is in the everyday that they rejoice and suffer; here and now."[7] However, Lefebvre also contends that everyday life, "a compound of insignificances united in this concept, responds and corresponds to modernity, a compound of signs by which our society expresses and justifies itself and which forms part of its ideology."[8] As Shields explains, for Lefebvre this idea of the "everyday" and of "everydayness" is thus not the same as simply everyday life or daily life; rather, the former refers to a more critical concept of an alienated life—one that we might considered to be socially dead.[9] Here, as Wander elaborates, the "'modern' world refers to the products of industrialization and the controls necessary to socialize workers and regulate consumption. It is a society whose rational character is defined through and has its limits set by highly organized groups (bureaucracies) operating through the state and/or the corporate state; whose level of operation takes place on and is aimed at influencing everyday life."[10]

Our everyday life is highly regimented, highly regulated, and highly disciplined. We work (normally) according to schedules imposed by others; the food we eat is (normally) inspected by regulatory agencies; we are required to obtain licenses and insurance when driving automobiles; and even what we read in magazines and view on the television screen is regulated to a certain degree. And while many of these government interventions into our day-to-day lives are exceptionally unremarkable, many others are not.

In *Everyday Life in the Modern World* Lefebvre writes of "terrorist societies" in which, on the one hand, "poverty and want [exist] and on the other a privileged class (possessing and administrating, exploiting, organizing and obtaining for its own ends as much social overtime as possible, either for ostentatious consumption or for accumulation, or indeed for both purposes at once) is maintained by the dual method of (ideological) *persuasion* and *compulsion* (punishment, laws and codes, courts, violence kept in store to prevent violence, overt violence, armed forces, police, etc.)." These "terrorist societies," in other words, emerge coincident with that of modern bureaucracies—with the modern state. And it is here that we see the transformation from "geographies of everyday life" to "geographies of everyday death": where the fears of being killed or being allowed to die become routine—Hannah Arendt's *banality of evil*. It is here, in these very modern spaces, that one sees how the "routine integration of terror into everyday life means that violence becomes increasingly normalized."[11]

We too often look away when confronted with intimate partner violence; we too often dismiss the "routine" bullying in the classroom as "child's play"; we dismiss the unmarked deaths of migrants seeking a better life in a foreign country; we look past the murder of homeless persons and prostitutes as . . . as what? As lives that do not matter?[12] Elizabeth Stanko documented the prevalence of just one set of violence—domestic violence—for just one day.[13] From midnight to midnight on September 28, 2000, a Thursday, Stanko audited every police service in the United Kingdom, all women's refuge and national helpline services, and any other relevant institution tasked with monitoring violence. Overall, Stanko estimates that an incident of domestic violence occurs in the UK every six to twenty seconds. She reports that the police received over 1,300 calls on that day for domestic violence (and more than 570,000 calls each year); that approximately 2,100 women and children were housed in refuges on that night—while another 200 women had asked for safe shelter, but were turned away because of a lack of space; and that nearly one in five counseling sessions of the other 900 sessions held that day at Relate, a national registered charity, mentioned domestic violence.[14] On that one "ordinary" day in the UK, violence was pervasive, yet unremarked. And what about for the United States? Japan? Any other country?

Is there a connection between our modern, violent societies and the practices of mass violence and genocide that typified the twentieth century? Is the unremarkable prevalence of violence indicative of something else? I suggest that the violence that pervades modern society is fundamentally different from the violence that typified societies that existed in the past. Modern violence has become routinized and normalized through state regulation. To be sure, rape and murder occurred hundreds of years ago; so too were people caught and punished for these crimes. But what is different is the relationship between "authority" and the "crimes." Whether a given act was even considered a crime has been modified by the emergence of the state. The same holds for our understanding—and practice—of mass violence: genocide.

The "concept" of genocide is a recent invention—but is the practice likewise "modern"? This is a question that has occupied the attention of genocide scholars throughout much of the late twentieth and early twenty-first centuries. As Martin Shaw explains, "Although some authorities have found much genocide in earlier history, others question whether systematically murderous policies towards ethnic and other groups were common."[15]

For Zygmunt Bauman, the Holocaust "was born and executed in our modern rational society, at the high stage of our civilization and at the peak of human cultural achievement, and for this reason it is a problem of that society, civilization and culture."[16] He continues that "the Holocaust was an outcome of a unique encounter between factors by themselves quite ordinary and common; and that the possibility of such an encounter could be blamed to a very large extent on the emancipation of the political state, with its monopoly of means of violence and its audacious engineering ambitions, from social control—following the step-by-step dismantling of all non-political power resources and institutions of social self-management."[17] In chapter 2 we saw how the Holocaust was composed of state-sanctioned and state-led violence that stemmed from a basic geographic imagination: to construct a pure living space for one population through the elimination of another population. We witnessed how sterilization practices were introduced to prevent future generations of lives considered to be unworthy of life; and how euthanasia was promoted to eliminate those unworthy lives presently living. We observed the concentration of populations for annihilation. Overall we saw the scientific management of life and the rational, bureaucratic legitimating of death.

In Maoist China, we saw a different geography of everyday death—not a bureaucracy intent on killing so that others may live, but rather a terrorist society that was prepared to let die millions of its own subjects to satisfy state goals. Mao Zedong's will to space was predicated on the discipline of bodies and the regulation of populations: to work harder, longer; to sacrifice more and expect less. Here, death became routine, as the Chinese state *intentionally* did nothing to prevent death in its attempt to produce a utopian political space. Likewise, in Cambodia, the Khmer Rouge legitimated death through the promotion of life: the life of a new state, a new society, and a new beginning. The death of those deemed unfit for the Khmer Rouge's imagined geography was no loss. The lives of all others were considered worthy only insofar as they could productively contribute to the political state. All were subject to the possibility of being killed or being allowed to "let die."

In total, the life and death experiences of Nazi Germany, Maoist China, and Democratic Kampuchea highlight that "genocide is neither capricious nor accidental."[18] Irving Horowitz is correct in his conclusion that "what made possible the engineering of death was a set of value-laden assumptions that the state, whether to purify its racial base or amplify its economic base, has the right to decide how many sacrifices are required to achieve its goals."[19] Leo Kuper concurs, writing that "the major arena for contemporary genocidal conflict and massacre is to be found within the sovereign state."[20] In other words, the key difference between earlier accounts of mass violence—such as the destruction of Carthage (149–146 BCE)—is the role of the modern state. Indeed, it bears repeating, as Martin Glassner notes, the "state" is such a familiar phenomenon that we tend to forget that only recently has all of the earth's land surface (except Antarctica) come under the jurisdiction and control of states.[21] This simple observation is of considerable importance and cannot be overemphasized. As the modern state gradually assumed form from the sixteenth century onwards, so too did our understanding of "population." Increasingly, government assumed the responsibility to manage the "naturalness" of births and deaths.

In myriad ways the state intervenes into matters of life and death. And it does so from that standpoint of trade-offs. How are scarce resources allocated? Who is eligible to receive benefits, and who is not? What is considered "homicide" or "manslaughter" or "acting in self-defense"? How are these decisions made; how are they justified, legitimated, and enforced? In Nazi Germany, we observed that the value of life and death was measured in sausages and jellies. But in many countries today, similar goals are made with respect to health care. Dranove, for example, observes that it is a laudable goal to spend whatever it takes to rid people of disease, suffering, and unnecessary dying. However, he maintains that this is "an unaffordable goal. If we do all we can to limit disease, prevent suffering, and prolong life, health care will claim the lion's share of our spending. There will be little if anything left to spend on justice, beauty, food, the national defense, or whatever else we hold dear. . . . [If] we are to achieve all our goals, we must be willing to curtail spending on health care. This means we must ration health care."[22]

These issues have been debated at length by both philosophers and medical ethicists. Concepts such as the "principle of non-vicious choice" have been forwarded, all in an attempt to provide a morally grounded calculation of life and death. John Harris, for example, begins with a simple scenario: two men are drowning and only one can be saved. Immediately, you are confronted with a life and death decision. How do you decide whom to save? Harris suggests that without *a priori* knowledge "it is a matter of moral indifference which we save."[23] In other words, while ideally we would like to save both people—that is not possible. And without knowing anything about the two people drowning, our choice is neutral; we have no moral reason to *prefer* to save one over the other. But then, Harris wonders, how a decision would be rendered if one of the drowning men were white and the other were black. At this point, can we expect that the bystander would be neutral? If the potential rescuer chose to save the *white* man because of racial prejudice, this would constitute "vicious discrimination." Harris explains, "It would be vicious not because the black man who was left to drown was entitled to be rescued *rather than* the white man, but because he was entitled not to be the victim of vicious discrimination."[24]

Harris's discussion is significant, in that it highlights how prejudices inform the calculated management of life. However, these decisions need not be so blatantly obvious. In Nazi Germany, Maoist China, and Democratic

Kampuchea, calculated decisions were made on a daily basis; many of these decisions were supported by appeals to science. As Horowitz concludes, the essential difference between the fourteenth and twentieth centuries is the "distinction between death as unavoidable tragedy willed by providence and death as manufactured purification of society willed by people."[25] It is therefore curious and significant that most scholars engaged in questions of biomedical ethics *mention* Nazi Germany, but quickly attempt to clarify that their decisions are in no way comparable to those of the Nazis. But are they so different?

Jonathan Glover concedes that the "idea of dividing people's lives into ones that are worth living and ones that are not is likely to seem both presumptuous and dangerous."[26] He acknowledges that apart from "seeming to indicate an arrogant willingness to pass godlike judgments on other people's lives, it may remind people of the Nazi policy of killing patients in mental hospitals."[27] Curiously, Glover maintains that "there is really nothing godlike in such a judgment. It is not a moral judgment we are making, if we think that someone's life is so empty and unhappy as to be not worth living."[28] This, of course, is the rationale employed by the Nazis.

Harris is also concerned about the Nazi experience. He wonders "if we accept that killing is permissible because it is a caring thing to do, because it saves lives, because it reduces suffering . . . in short, because it is right, how can we be sure that it will only be done when it is right and when it does achieve all these things?"[29] Harris warns, "Once accepted, might not such killing become more and more common and the horror of ending the lives of others less and less real, until something approaching a Nazi callousness became pervasive?"[30]

Both Glover and Harris are concerned about the "slippery slope" and what Tony Hope calls "playing the Nazi card."[31] For Hope, comparisons with Nazi policy and practice are misleading and inappropriate for present debates on matters of life and death: abortion, euthanasia, and so on. Hope states that "what the Nazis did was to kill people without any consideration of benefit to the person killed."[32] Hope is wrong on this count. As the historical scholarship of Robert Lifton, Michael Burleigh, and Robert Proctor documents, Germany's leading medical practitioners, anthropologists, geneticists, and ethicists *believed* in the actions called for by the Nazi Party. These individuals, coupled with lawyers, engineers, and untold other professionals, maintained that the most advanced science supported the calculated management of life and death. Moreover, they *believed* that some lives were not worth living, and that the elimination of those lives was an act of mercy. From the Nazi point of view, even those slated for euthanasia would understand the rationale for their own deaths.

Hope also introduces the concept of the "slippery slope" argument in the context of the Nazi experience. The slippery slope argument holds that "once you accept one particular position [such as allowing euthanasia of terminally ill patients] then it will be extremely difficult, or indeed impossible, not to accept more and more extreme positions."[33] He explains that the "main counter to the slippery slope argument is to claim that a barrier can be placed part way down the slope that in stepping onto the top of the slope we will not inevitably slide to the bottom—but only as far as the barrier."[34] Harris, again, uses the Nazi example in his discussion of the slippery slope. I quote Harris at length:

> The Nazi euthanasia programme was nothing like the possibilities we are considering. Under the Nazis euthanasia was simply one way of exterminating those racially or politically beyond moral consideration. And the Nazis were not short of other ways to achieve the same ends. It is precisely because we care about spina bifida children, precisely because we are in no doubt that they must not suffer, that we are concerned about what it is in their best interests to do. The spectre of Nazism offers no analogy at all and so only fogs the issues.[35]

Harris's refusal to engage seriously with the Nazi program is surprising to say the least. He seems to ignore, again, that the Nazi medical establishment likewise evinced "care" of the mentally or physically disabled. And more significant, he ignores that the Nazi bureaucracy engaged in a supposedly rational calculated management of life—just as Dranove and other contemporary writers suggest for our approach to health care. We cannot afford to save *all lives*; hard decisions must be made. And as the subtitle of Dranove's book so effectively asks: "Who Lives? Who Dies? And Who Decides?"

An engagement with Nazi Germany, Maoist China, and Democratic Kampuchea does not "fog the issue." It does highlight some difficult and uncomfortable comparisons. How does the state—broadly conceived—manage the everydayness of life and death? When one considers the state's *will to space*, the production of sovereign political space, who is to be included or excluded from participation? What is the place of women in society? Of ethnic minorities? Of noncitizens? Who is denied access to health care—either because of poverty or because of insufficient resources? How does the state promote or discourage reproduction? Are all forms of contraception readily available? What of abortion? Or surrogate motherhood? How is death determined and what are the implications? For organ donation? For capital punishment? For better or for worse, these questions were discussed, debated, and acted upon in Nazi Germany. And in Maoist China. And in Democratic Kampuchea. Far from fogging the issue, these episodes of mass violence speak directly to the spatiality of life and death—both then and now. As Horowitz writes, the fundamental unit for taking or preserving life becomes the state.[36]

And what of the state in the twenty-first century? Nikolas Rose, for example, argues that "life may, today more than ever, be subject to judgments of value, but those judgments are not made by a state managing the population en masse."[37] Rose sees a reduced function of the sovereign state, suggesting that the "state is no longer expected to resolve society's needs for health" and that the "vitality of the species—the nation, the population, the race—is rarely the rational and legitimation for compulsory interventions into the individual lives of those who are only its constituent elements."[38] Rose instead points to "a whole range of pressure groups, campaigning organizations, and self-help groups" that "have come to occupy the space of desires, anxieties, disappointments, and ailments between the will to health and the experience of its absence."[39]

I would not be so quick to rule out the salience of state intervention into matters of life and death—witness the political battles over abortion, euthanasia, stem-cell research, capital punishment—even same-sex marriage legislation. Within the United States, as a case in point, we see daily how these issues become electoral issues, influencing political campaigns both local and national. Moreover, we see in the United States how Congress attempts to legislate these matters; and the Supreme Court determines their constitutionality. The "state" provides the context in which pressure groups, campaign organizations—even doctors—must operate.

Rose likewise fails to appreciate the continuity between the Holocaust and contemporary society. In his critique of both Michel Foucault and Giorgio Agamben, for example, Rose asserts that "while the lives, illnesses, and troubles of many may be ignored or marginalized in contemporary political economies of vitality, to let die is not to make die—no 'sovereign' wills or plans the sickness and death of our fellow citizens."[40] This statement is interesting, given its many omissions. Can we not identify any practice by which an "advanced liberal polity" (as Rose discusses) plans the death of a fellow citizen? Capital punishment readily comes to mind. The United States is one of the few remaining Western countries in which *the state plans the death of its fellow citizens.*[41] We might also consider the extrajudicial use of political assassination—a practice that is illegal but nevertheless employed by the United States.

And what of the notion that "to let die is not to make die"? This is a contentious point—as reflected by the many writings of Jonathan Glover, John Harris, James Rachels, Tony Hope, Jeff McMahan, and so on. For some philosophers and bioethicists, there is no moral difference between "letting die" and killing, while for others, presumably Rose, there is a strong moral difference. My reading of Germany, China, and Cambodia suggests that the state, one way or the other, will establish its policies and practices in line with this distinction.

Rose is correct—and this is a point demanding more critical work—in identifying the influence of nonstate actors (e.g., pharmaceutical companies). However, such work requires a positioning of nonstate actors within the context of state intervention. It is the state that determines which industries are regulated—and to what extent these are regulated. It is the state that provides the context for medical licensing; and it is the state that crafts the nation's overall health care system.

I began this book with a concern over the value of life. I suggested that, conceptually, we might approach genocide and mass violence as a "will to space." My thesis is straightforward: the social processes leading to moral inclusion or exclusion have a geographic component. When states intervene into our daily lives, when states regulate those processes affecting life and death, when states determine who may or may not occupy their sovereign territory, they do so from the standpoint of inclusion and exclusion. Accordingly, geographies of everyday life and death become sites of contestation, the locus of sociospatial control where ideologies of racism, sexism, classism, ablism, and so forth are enacted. The decisions to kill (as in capital punishment or euthanasia), to foster life (through the provision of health care and medical resources), or to disallow life to the point of death (through access to health insurance) are made within a context of state valuation. It is the modern, bureaucratic sovereign state that oversees the legalities of sterilization, contraceptive techniques, abortion, capital punishment, health-care rationing and so many other areas of our day-to-day lives. As Agamben states quite clearly: "In modern biopolitics, sovereign is he who decides on the value or the nonvalue of life as such."[42]

NOTES

1. Joe Painter, "Prosaic Geographies of Stateness," *Political Geography* 25(2006): 752–74; at 753.

2. Ben Highmore, *Everyday Life and Cultural Theory: An Introduction* (New York: Routledge, 2002), 1.

3. Philip Wander, "Introduction" in Henri Lefebvre, *Everyday Life in the Modern World* (New Brunswick, NJ: Transaction Publishers, 2002), vii–xxiii; at vii–viii.

4. Joshua Inwood, "Making the Legal Visible: Wilhelmina Griffin Jones' Experience of Living in Alabama during Segregation," *Southeastern Geographer* 45(2005): 54–66; at 60.

5. David Delaney, *Race, Place, and the Law, 1836–1948* (Austin: University of Texas Press, 1998), 4.

6. Wander, "Introduction," viii.

7. Henri Lefebvre, *Everyday Life in the Modern World* (New Brunswick, NJ: Transaction Publishers, 2002), 21.

8. Lefebvre, *Everyday Life*, 24.

9. Rob Shields, *Lefebvre, Love and Struggle* (New York: Routledge, 1998), 66.

10. Wander, "Introduction," viii.

11. Allen Feldman, "X-Children and the Militarisation of Everyday Life: Comparative Comments on the Politics of Youth, Victimage and Violence in Transitional Societies," *International Journal of Social Welfare* 11(2002): 286–99; at 294.

12. James A. Tyner, *Space, Place, and Violence: Violence and the Embodied Geographies of Race, Sex, and Gender* (New York: Routledge, 2011).

13. Elizabeth A. Stanko, "The Day to Count: Reflections on a Methodology to Raise Awareness about the Impact of Domestic Violence in the UK," *Criminology and Criminal Justice* 1(2001): 215–26.

14. Stanko, "The Day to Count," 224

15. Martin Shaw, *What Is Genocide?* (Cambridge, UK: Polity Press, 2007), 133.

16. Zygmunt Bauman, *Modernity and the Holocaust* (Ithaca, NY: Cornell University Press, 1989), x.

17. Bauman, *Modernity,* xiii.

18. Irving Louis Horowitz, *Taking Lives: Genocide and State Power*, 3rd ed. (New Brunswick, NJ: Transaction Books, 1980), xi.

19. Horowitz, *Taking Lives*, 6.

20. Leo Kuper, *Genocide: Its Political Use in the Twentieth Century* (New Haven, CT: Yale University Press, 1981), 14.

21. Martin Ira Glassner, *Political Geography* 2nd ed. (New York: John Wiley & Sons, 1996), 53.

22. David Dranove, *What's Your Life Worth? Health Care Rationing . . . Who Lives? Who Dies? And Who Decides?* (Upper Saddle River, NJ: Prentice Hall, 2003), 2.

23. John Harris, *The Value of Life: An Introduction to Medical Ethics* (New York: Routledge, 1985), 70.

24. Harris, *The Value of Life*, 71.

25. Horowitz, *Taking Lives*, 6.

26. Jonathan Glover, *Causing Death and Saving Lives* (New York: Penguin Books, 1977), 52.

27. Glover, *Causing Death*, 52.

28. Glover, *Causing Death*, 52.

29. Harris, *The Value of Life*, 81–82.

30. Harris, *The Value of Life*, 82.

31. Tony Hope, *Medical Ethics: A Very Short Introduction* (Oxford: Oxford University Press, 2004), 8.

32. Hope, *Medical Ethics*, 10.

33. Hope, *Medical Ethics*, 70.

34. Hope, *Medical Ethics*, 70.

35. Harris, *The Value of Life*, 36.

36. Horowitz, *Taking Lives*, 35.

37. Nikolas Rose, *The Politics of Life Itself: Biomedicine, Power, and Subjectivity in the Twenty-First Century* (Princeton: Princeton University Press, 2007), 58.

38. Rose, *Politics of Life*, 63.

39. Rose, *Politics of Life*, 64.

40. Rose, *Politics of Life*, 58.

41. Terance D. Miethe and Hong Lu, *Punishment: A Comparative Historical Perspective* (Cambridge, MA: Cambridge University Press, 2005).

42. Agamben, *Homo Sacer*, 142.

Selected Bibliography

Admiraal, Pieter. "Euthanasia and Assisted Suicide," in *Birth to Death: Science and Bioethics*, edited by David C. Thomasma and Thomasine Kushner (Cambridge: Cambridge University Press, 1996), 207–17.

Agamben, Giorgio. *Homo Sacer: Sovereign Power and Bare Life* (Stanford, CA: Stanford University Press, 1998).

———. *Remnants of Auschwitz: The Witness and the Archive* (New York: Zone Books, 2002).

Agnew, John. "The Territorial Trap: The Geographical Assumptions of International Relations Theory," *Review of International Political Economy* 1(1994): 53–80.

Aitken, Stuart C. *Family Fantasies and Community Space* (New Brunswick, NJ: Rutgers University Press, 1998).

Arad, Yitzhak. *Belzec, Sobibor, Treblinka: The Operation Reinhard Death Camps* (Bloomington: Indiana University Press, 1999).

Arendt, Hannah. *On Violence* (New York: Harcourt, 1970).

Ashcroft, Bill, and Pal Ahluwalia. *Edward Said* (New York: Routledge, 1999).

Ayers, David M. *Anatomy of a Crisis: Education, Development, and the State in Cambodia, 1953–1998* (Chiang Mai, Thailand: Silkworm Press, 2000).

Baker, L. H. M. *Race Improvement or Eugenics* (New York, n.p., 1912).

Barker, Philip. *Michel Foucault: An Introduction* (Edinburgh: Edinburgh University Press, 1998).

Bartky, Sandra Lee. *Femininity and Domination: Studies in the Phenomenology of Oppression* (New York: Routledge, 1990).

Bassin, Mark. "Imperialism and the Nation State in Friedrich Ratzel's Political Geography," *Progress in Human Geography* 11(1987): 473–95.

Bauman, Zygmunt. *Modernity and the Holocaust* (Ithaca, NY: Cornell University Press, 2000).

Becker, Jasper. *Hungry Ghosts: Mao's Secret Famine* (New York: Henry Holt & Co., 1998).

Bergen, Doris. *The Holocaust: A Concise History* (Lanham: Rowman & Littlefield, 2009).

Bertov, Omer (ed.). *The Holocaust: Origins, Implementations, Aftermath* (New York: Routledge, 2000).

Browning, Christopher. *Nazi Policy, Jewish Workers, German Killers* (Cambridge: Cambridge University Press, 2000).

———. *Ordinary Men: Reserve Police Battalion 101 and the Final Solution in Poland* (New York: Harper Perennial, 1992).

Burleigh, Michael. *Death and Deliverance: "Euthanasia" in Germany 1900–1945* (Cambridge: Cambridge University Press, 1994).

———. *Moral Combat: Good and Evil in World War II* (New York: Harper, 2011).

Burleigh, Michael, and Wolfgang Wippermann. *The Racial State: Germany 1933–1945* (Cambridge: Cambridge University Press, 1991).

Cahill, George. "Famine Symposium: Physiology of Acute Starvation in Man," *Ecology of Food and Nutrition* 6(1978): 221–30.

Caldwell, John. "Demographers and the Study of Mortality: Scope, Perspectives, and Theory," *Annals of the New York Academy of Sciences* 954(2001): 19–34.

Cartwright, Will. "Killing and Letting Die: A Defensible Distinction," *British Medical Bulletin* 52(1996): 354–61.

Chalk, Frank, and Kurt Jonassohn. *The History and Sociology of Genocide: Analyses and Case Studies* (New Haven, CT: Yale University Press, 1990).

Chandler, David P. *The Tragedy of Cambodian History: Politics, War, and Revolution since 1945* (New Haven, CT: Yale University Press, 1991).

———. *Voices from S-21: Terror and History in Pol Pot's Secret Prison* (Berkeley: University of California Press, 1999).

———. *A History of Cambodia*, 3rd edition (Boulder, CO: Westview Press, 2000).

———. *Brother Number One: A Political Biography of Pol Pot*, revised edition (Chiang Mai, Thailand: Silkworm Press, 2000).

Chang, Gene Hsin, and Guanzhong James Wen. "Food Availability Versus Consumption Efficiency: Causes of the Chinese Famine," *China Economic Review* 9(1998): 157–66.

Chang, Jung, and Jon Halliday. *Mao: The Unknown Story* (New York: Anchor Books, 2006).

Chirot, Daniel, and Clark McCauley. *Why Not Kill Them All? The Logic and Prevention of Mass Political Murder* (Princeton, NJ: Princeton University Press, 2006).

Clarke, David B., Marcus A. Doel, and Francis X. McDonough. "Holocaust Topologies: Singularlity, Politics, Space," *Political Geography* 15(1996): 457–89.

Clayton, Thomas. "Building the New Cambodia: Educational Destruction and Construction under the Khmer Rouge, 1975–1979," *History of Education Quarterly* 38(1998): 1–16.

———. "Re-Orientations in Moral Education in Cambodia since 1975," *Journal of Moral Education* 34(2005): 505–17.

Commons, John R. *Races and Immigrants in America* (New York: Macmillan, 1907).

Crampton, Jeremy, and Stuart Elden. "Space, Politics, Calculation: An Introduction," *Social and Cultural Geography* 7(2006): 681–85.

Cresswell, Tim. *In Place/Out of Place: Geography, Ideology and Transgression* (Minneapolis: University of Minnesota Press, 1996).

Curtis, Bruce. "Foucault on Governmentality and Population: The Impossible Discovery," *Canadian Journal of Sociology* 27(2002): 505–33.

Dahlman, Carl T. "Sovereignty," in *Key Concepts in Political Geography*, edited by C. Gallaher, C. T. Dahlman, M. Gilmartin, A. Mountz, and P. Shirlow (Los Angeles: Sage Publications, 2009), 28–40.

Daley, Martin, and Margo Wilson. *Homicide* (London: Transaction Publishers, 1988).

Danielsson, Sarah K. "Creating Genocidal Space: Geographers and the Discourse of Annihilation, 1880–1933," *Space and Polity* 13(2009): 55–68.

Davenport, Charles B. *Eugenics: The Science of Human Improvement by Better Breeding* (New York: Henry Holt, 1910).

Dean, Mitchell. *Governmentality: Power and Rule in Modern Society* (Thousand Oaks, CA: Sage Publications, 1999).

Delaney, David. *Race, Place, and the Law, 1836–1948* (Austin: University of Texas Press, 1998).

De Waal, Frans. *Chimpanzee Politics: Power and Sex among Apes* (Baltimore: Johns Hopkins Press, 1992).

———. *Our Inner Ape: A Leading Primatologist Explains Why We Are Who We Are* (New York: Riverhead Books, 2005).

De Walque, Damien. "Selective Mortality during the Khmer Rouge Period in Cambodia," *Population and Development* 31(2005): 351–68.

———. "The Socio-Demographic Legacy of the Khmer Rouge Period in Cambodia," *Population Studies* 60(2006): 223–31.

Dikötter, Frank. *Mao's Great Famine: The History of China's Most Devastating Catastrophe, 1958–1962* (New York: Walker & Co., 2010).

Dirks, Robert. "Social Responses during Severe Food Shortages and Famine," *Current Anthropology* 21(1980): 21–44.

Dranove, David. *What's Your Life Worth? Health Care Rationing . . . Who Lives? Who Dies? And Who Decides?* (New York: Prentice Hall, 2003).

Dreyfus, Hubert L., and Paul Rabinow. *Michel Foucault: Beyond Structuralism and Hermeneutics*, 2nd edition (Chicago: University of Chicago Press, 1983).

Dubow, Saul. *Scientific Racism in Modern South Africa* (Cambridge: Cambridge University Press, 1995).

Dy, Khamboly. *A History of Democratic Kampuchea (1975–1979)* (Phnom Penh: Documentation Center of Cambodia, 2007).

Ebihara, May. "Revolution and Reformulation in Kampuchean Village Culture," in *The Cambodian Agony*, 2nd edition, edited by David A. Ablin and Marlowe Hood (New York: M.E. Sharpe, 1990), 16–61.

Edelman, Murray. *Political Language: Words That Succeed and Policies That Fail* (New York: Academic Press, 1977).

Elden, Stuart. "Contingent Sovereignty, Territorial Integrity and the Sanctity of Borders," *SAIS Review* 26(2006): 11–25.

Em, Sokhym. "Female Patients," *Searching for the Truth* 33(2002): 25–29.

———. "Rabbit Dropping Medicine," *Searching for the Truth* 30(2002): 22–23.

——. "Revolutionary Female Medical Staff in Tram Kak District, Part I," *Searching for the Truth* 34(2002): 24–27.

——. "Revolutionary Female Medical Staff in Tram Kak District, Part II," *Searching for the Truth* 35(2002): 17–19.

Esposito, Roberto. *Bíos: Biopolitics and Philosophy*, translated by Timothy Campbell (Minneapolis: University of Minnesota Press, 2008).

Etcheson, Craig. *After the Killing Fields: Lessons from the Cambodia Genocide* (Lubbock: Texas Tech University Press, 2005).

Fein, Helen. *Imperial Crime and Punishment: The Massacre at Jallianwala Bagh and British Judgment, 1919–1920* (Honolulu: University of Hawaii Press, 1977).

Feldman, Allen. "X-Children and the Militarisation of Everyday Life: Comparative Comments on the Politics of Youth, Victimage and Violence in Transitional Societies," *International Journal of Social Welfare* 11(2002): 286–99.

Foucault, Michel. *Discipline and Punish: The Birth of the Prison*, translated by Alan Sheridan (New York: Vintage Books, 1979).

——. "Two Lectures," in *Power/Knowledge: Selected Interviews and Other Writings, 1972–1977*, edited by C. Gordon (New York: Pantheon Books, 1980), 78–108.

——. *The History of Sexuality: An Introduction* (New York: Vintage Books, 1990).

——. "The Political Technology of Individuals," in *Power: Essential Works of Foucault, 1954–1984*, volume 3, edited by Paul Rabinow (New York: New Press, 2000), 403–17.

——. "The Subject and Power," in *Power: Essential Works of Foucault, 1954–1984*, volume 3, edited by Paul Rabinow (New York: New Press, 2000), 326–48.

——. *"Society Must Be Defended": Lectures at the Collège de France, 1975–1976*, translated by David Macey (New York: Picador, 2003).

——. *Security, Territory, Population: Lectures at the Collège de France, 1977–1978*, translated by Graham Burchell (New York: Picador, 2007).

Friedlander, Henry. *The Origins of Nazi Genocide: From Euthanasia to the Final Solution* (Chapel Hill: University of North Carolina Press, 1995).

Fritzsche, Peter. *Life and Death in the Third Reich* (Cambridge, MA: Belknap Press of Harvard University Press, 2008).

Gamson, William A. "Hiroshima, the Holocaust, and the Politics of Exclusion," *American Sociological Review* 60(1995): 1–20.

Gilligan, James. *Violence: Reflections on a National Epidemic* (New York: Vintage Books, 1997).

Glassner, Martin. *Political Geography*, 2nd edition (New York: John Wiley & Sons, 1996).

Glover, Jonathan. *Causing Death and Saving Lives* (New York: Penguin, 1990 [1977]).

Green, O. H. "Killing and Letting Die," *American Philosophical Quarterly* 17(1980): 195–204.

Gregory, Derek. *Geographical Imaginations* (Cambridge, MA: Blackwell, 1994).

——. *The Colonial Present: Afghanistan, Palestine, Iraq* (Cambridge, MA: Blackwell, 2004).

———. "The Lightning of Possible Storms," *Antipode* 36(2004): 798–808.

Grossman, Dave. *On Killing: The Psychological Cost of Learning to Kill in War and Society* (New York: Back Bay Books, 1996).

Haller, Jr., J. S. *Outcasts from Evolution: Scientific Attitudes of Racial Inferiority, 1859–1900* (Carbondale: Southern Illinois University Press, 1971).

Hamburg, David. *Preventing Genocide: Practical Steps Toward Early Detection and Effective Action* (Boulder, CO: Paradigm Publishers, 2008).

Harris, John. *The Value of Life: An Introduction to Medical Ethics* (New York: Routledge, 1985).

Heuveline, Patrick. "'Between One and Three Million': Towards the Demographic Reconstruction of a Decade of Cambodian History," *Population Studies* 52(1998): 49–65.

Highmore, Ben. *Everyday Life and Cultural Theory: An Introduction* (New York: Routledge, 2002).

Hilberg, Raul. *The Destruction of the European Jews* (Chicago: Quadrangle, 1961).

Hinton, Alexander L. *Why Did They Kill? Cambodia in the Shadow of Genocide* (Berkeley: University of California Press, 2005).

Hope, Tony. *Medical Ethics: A Very Short Introduction* (Oxford: Oxford University Press, 2004).

Horowitz, Irving Louis. *Taking Lives: Genocide and State Power*, 3rd edition (New Brunswick, NJ: Transaction Books, 1980).

Huckle, J. "Geography and Schooling," in *The Future of Geography*, edited by Ron Johnston (London: Methuen, 1985), 291–306.

Hutchings, Graham. *Modern China: A Guide to a Century of Change* (Cambridge, MA: Harvard University Press, 2008).

Inwood, Joshua. "Making the Legal Visible: Wilhelmina Griffin Jones' Experience of Living in Alabama during Segregation," *Southeastern Geographer* 45(2005): 54–66.

Jackson, Karl D. "The Ideology of Total Revolution," in *Cambodia 1975–1978: Rendezvous with Death*, edited by Karl D. Jackson (Princeton, NJ: Princeton University Press, 1989), 37–78.

Jackson, Robert. *Sovereignty: Evolution of an Idea* (Malden, MA: Polity Press, 2007).

James, Joy. *Resisting State Violence: Radicalism, Gender and Race in U.S. Culture* (Minneapolis: University of Minnesota Press, 1996).

James, Preston E., and Geoffrey J. Martin. *All Possible Worlds: A History of Geographical Ideas*, 2nd edition (New York: John Wiley & Sons, 1981).

Johnson, D. Gale. "China's Great Famine: Introductory Remarks," *China Economic Review* 9(1998): 103–9.

Kahn, Paul W. *Sacred Violence: Torture, Terror, and Sovereignty* (Ann Arbor: University of Michigan Press, 2008).

Kashani-Sabet, F. "Picturing the Homeland: Geography and National Identity in Late Nineteenth- and Early Twentieth-Century Iran," *Journal of Historical Geography* 24(1998): 413–30.

Kevles, Daniel J. *In the Name of Eugenics: Genetics and the Uses of Human Heredity* (Berkeley: University of California Press, 1985).

Keys, Ancel, Josef Brozek, Austin Henschel, Olaf Mickelson, and Henry Taylor. *The Biology of Human Starvation* (Minneapolis: University of Minnesota Press, 1950).

Kiernan, Ben. *How Pol Pot Came to Power: A History of Communism in Kampuchea, 1930–1975* (London: Verso, 1985).

———. "The Genocide in Cambodia, 1975–1979," *Bulletin of Concerned Asian Scholars* 22(1990): 35–40.

———. *The Pol Pot Regime: Policies, Race and Genocide in Cambodia under the Khmer Rouge, 1975–1979* (New Haven, CT: Yale University Press, 1996).

———. "The Demography of Genocide in Southeast Asia: The Death Tolls in Cambodia, 1975–1979, and East Timor, 1975–1980," *Critical Asian Studies* 35(2003): 585–97.

———. *Blood and Soil: A World History of Genocide and Extermination from Sparta to Darfur* (New Haven, CT: Yale University Press, 2007).

———. "The Cambodian Genocide, 1975–1979," in *Century of Genocide: Critical Essays and Eyewitness Accounts*, 3rd edition, edited by Samuel Totten and W. S. Parsons (New York: Routledge, 2009), 341–75.

Koenigsberg, Richard A. *Nations Have the Right to Kill: Hitler, the Holocaust and War* (New York: Library of Social Sciences, 2009).

Koonz, Claudia. *The Nazi Conscience* (Cambridge, MA: Belknap Press of Harvard University Press, 2003).

Krasner, S. D. "Problematic Sovereignty," in *Problematic Sovereignty: Contested Rules and Political Possibilities*, edited by S. D. Krasner (New York: Columbia University Press, 2001), 1–23.

Kühl, Stefan. *The Nazi Connection: Eugenics, American Racism, and German National Socialism* (Oxford: Oxford University Press, 1994).

Kuper, Leo. *Genocide: Its Political Use in the Twentieth Century* (New Haven, CT: Yale University Press, 1981).

Larson, Edward J. *Sex, Race, and Science: Eugenics in the Deep South* (Baltimore: Johns Hopkins University Press, 1995).

Lefebvre, Henri. *The Production of Space*, translated by D. Nicholson-Smith (Oxford, UK: Blackwell, 1991).

———. *Everyday Life in the Modern World* (New Brunswick, NJ: Transaction Publishers, 2002).

Legg, Stephen. "Foucault's Population Geographies: Classifications, Biopolitics and Governmental Spaces," *Population, Space and Place* 11(2005): 137–56.

Leitenberg, Milton. "Deaths in Wars and Conflicts in the 20th Century," *Cornell University Peace Studies Program Occasional Paper #29* (Ithaca, NY: Cornell University Press).

Lemke, Thomas. *Biopolitics: An Advanced Introduction*, translated by Eric F. Trump (New York: New York University Press, 2011).

Lerner, Melvin J. *The Belief in a Just World: A Fundamental Decision* (New York: Plenum Press, 1980).

Levene, Mark. "Why is the Twentieth Century the Century of Genocide?" *Journal of World History* 11(2000): 305–36.

Levi, Primo. *The Drowned and the Saved* (New York: Vintage Books, 1989).

LeVine, Peg. *Love and Dread in Cambodia: Weddings, Births, and Ritual Harm under the Khmer Rouge* (Singapore: National University of Singapore Press, 2010).

Lewis, Marin, and Kären Wigen. *The Myth of Continents: A Critique of Metageography* (Berkeley: University of California Press, 1997).

Leyton, G. B. "Effects of Slow Starvation," *Lancet* 2(1946): 73–79.

Lifton, Robert J. *The Nazi Doctors: Medical Killing and the Psychology of Genocide* (New York: Basic Books, 1986).

Lin, Justin Yifu, and Dennis Tao Yang. "On the Causes of China's Agricultural Crisis and the Great Leap Famine," *China Economic Review* 9(1998): 125–40.

Livingstone, David N. *The Geographical Tradition: Episodes in the History of a Contested Enterprise* (Cambridge, MA: Basil Blackwell, 1992).

Locard, Henri. *Pol Pot's Little Red Book: The Sayings of Angkar* (Chiang Mai, Thailand: Silkworm Press, 2004).

Mam, Kalyanee. "The Endurance of the Cambodian Family under the Khmer Rouge Regime: An Oral History," in *Genocide in Cambodia and Rwanda: New Perspectives*, edited by Susan E. Cook (New Haven, CT: Yale Center for International and Area Studies, Genocide Studies Program Monograph Series, no. 1, 2004), 127–71.

Mam, Teeda Butt. "Worms from Our Skin," in *Children of Cambodia's Killing Fields: Memoirs by Survivors*, compiled by Dith Pran and edited by Kim DePaul (New Haven, CT: Yale University Press, 1997), 11–17.

Mamadouh, Virginie. "Geography of War, Geographers and Peace," in *Geography of War and Peace: From Death Camps to Diplomats*, edited by Colin Flint (Oxford: Oxford University Press, 2005), 26–60.

Marsden, R. "A Political Technology of the Body: How Labour Is Organized into a Productive Force," *Critical Perspectives on Accounting* 9(1998): 99–136.

Marshall, S. L. A. *Men Against Fire* (Gloucester: Peter Smith, 1978).

Mbembe, Achilles. "Necropolitics," *Public Culture* 15(2003): 11–40.

McDowell, Linda. *Gender, Identity & Place: Understanding Feminist Geographies* (Minneapolis: University of Minnesota Press, 1999).

McKale, Donald M. *Hitler's Shadow War: The Holocaust and World War II* (Lanham: Taylor Trade, 2002).

McMahan, Jeff. *The Ethics of Killing: Problems at the Margins of Life* (New York: Oxford University Press, 2002).

Miethe, Terance, and Hong Lu. *Punishment: A Comparative Historical Perspective* (Cambridge: Cambridge University Press, 2005).

Miles, Robert. *Racism* (New York: Routledge, 1989).

Morgan, J. "Imagined Country: National Environmental Ideologies in School Geography Textbooks," *Antipode* 35(2003): 444–62.

Mosse, George L. *Toward the Final Solution: A History of European Racism* (New York: Harper & Row, 1978).

Ngor, Haing (with R. Warner). *Survival in the Killing Fields* (New York: Carroll & Graf Publishers, 1987).

Opotow, Susan. "Reconciliation in Time of Impunity: Challenges for Social Justice," *Social Justice Research* 14(2001): 149–70.

Ordover, Nancy. *American Eugenics: Race, Queer Anatomy, and the Science of Nationalism* (Minneapolis: University of Minnesota Press, 2003).

Ó Tuathail, Gearóid. *Critical Geopolitics* (Minneapolis: University of Minnesota Press, 1996).

Oveson, Jan, and Ing-Britt Trankell. *Cambodians and Their Doctors: A Medical Anthropology of Colonial and Post-Colonial Cambodia* (Copenhagen: Nordic Institute of Asian Studies, 2010).

Painter, Joe. "Prosaic Geographies of Stateness," *Political Geography* 25(2006): 752–74.

Patterson, Orlando. *Slavery and Social Death: A Comparative Study* (Cambridge, MA: Harvard University Press, 1982).

Paul, Diane B. *Controlling Human Heredity: 1865 to the Present* (New York: Macmillan, 1995).

Philo, Chris. "Sex, Life, Death, Geography: Fragmentary Remarks Inspired by Foucault's Population Geographies," *Population, Space and Place* 11(2005): 325–33.

Pink, Thomas. *Free Will: A Very Short Introduction* (Oxford: Oxford University Press, 2004).

Popenoe, Paul B., and Roswell Hill Johnson. *Applied Eugenics*, 2nd edition (New York: Macmillan, 1918).

Porter, Gareth. "Vietnamese Communist Policy toward Kampuchea, 1930–1970," in *Revolution and Its Aftermath: Eight Essays*, edited by David P. Chandler and Ben Kiernan (New Haven, CT: Yale University Southeast Asian Studies, 1983), 57–98.

Pouchad, François. "Social Change in the Vortex of Revolution," in *Cambodia 1975–1978: Rendezvous with Death*, edited by Karl D. Jackson (Princeton, NJ: Princeton University Press, 1989), 151–77.

Prak, Sarom. "The Unfortunate Cambodia," in *Children of Cambodia's Killing Fields: Memoirs by Survivors*, compiled by Dith Pran and edited by Kim DePaul (New Haven, CT: Yale University Press, 1997), 67–71.

Pringle, Heather. *The Master Plan: Himmler's Scholars and the Holocaust* (New York: Hyperion, 2006).

Proctor, Robert N. *Racial Hygiene: Medicine under the Nazis* (Cambridge, MA: Harvard University Press, 1988).

Quinn, Kenneth M. "The Pattern and Scope of Violence," in *Cambodia 1975–1978: Rendezvous with Death*, edited by Karl D. Jackson (Princeton, NJ: Princeton University Press, 1989), 179–208.

Rabinow, Paul, and Nikolas Rose. "Biopower Today," *BioSocieties* 1(2006): 195–217.

Rachels, James. "Killing and Starving to Death," *Philosophy* 54(1979): 159–71.

Rees, Laurence. *Auschwitz: A New History* (New York: Public Affairs, 2005).

Rhodes, Richard. *Masters of Death: The SS-Einsatzgruppen and the Invention of the Holocaust* (New York: Vintage Books, 2003).

Ridley, Matt. *The Red Queen: Sex and the Evolution of Human Nature* (New York: Harper Perennial, 2003).

Riskin, Carl. "Seven Questions about the Chinese Famine of 1959–61," *China Economic Review* 9(1998): 111–24.

Rose, Nikolas. *The Politics of Life Itself: Biomedicine, Power, and Subjectivity in the Twenty-First Century* (Princeton, NJ: Princeton University Press, 2007).

Rummel, Rudolph J. *Death by Government: Genocide and Mass Murder in the Twentieth Century* (New Brunswick, NJ: Transaction Publishers, 1994).

Said, Edward. *Orientalism* (New York: Vintage Books, 1979).

Sam, Roeun. "Living in Darkness," in *Children of Cambodia's Killing Fields: Memoirs by Survivors*, compiled by Dith Pran and edited by Kim DePaul (New Haven, CT: Yale University Press, 1997), 73–81.

Sarat Austin, and Jennifer L. Culbert. "Introduction: Interpreting the Violence State," in *States of Violence: War, Capital Punishment, and Letting Die*, edited by Austin Sarat and Jennifer L. Culbert (Cambridge: Cambridge University Press, 2009), 1–22.

Sarup, Madan. *An Introductory Guide to Post-Structuralism and Postmodernism*, 2nd edition (Athens: University of Georgia Press, 1993).

Schulten, Susan. *The Geographical Imagination in America, 1880–1950* (Chicago: University of Chicago Press, 2001).

Semelin, Jacques. *Purify and Destroy: The Political Uses of Massacre and Genocide*, translated by Cynthia Schoch (New York: Columbia University Press, 2007).

Sen, Amartya. *Identity and Violence: The Illusion of Destiny* (New York: W.W. Norton, 2006).

Shapiro, Judith. *Mao's War against Nature: Politics and the Environment in Revolutionary China* (Cambridge: Cambridge University Press, 2001).

Shaw, Martin. *War & Genocide: Organized Killing in Modern Society* (Malden, MA: Polity Press, 2003).

———. *What Is Genocide?* (Cambridge, UK: Polity Press, 2007).

Shields, Rob. *Lefebvre, Love and Struggle* (New York: Routledge, 1998).

Short, Philip. *Pol Pot: Anatomy of a Nightmare* (New York: Henry Holt & Co., 2004).

Sibley, David. *Geographies of Exclusion: Society and Difference in the West* (London: Routledge, 1995).

Smith, David Livingstone. *The Most Dangerous Animal: Human Nature and the Origins of War* (New York: St. Martin's Griffin, 2007).

Soja, Edward W. *Postmodern Geographies: The Reassertion of Space in Critical Social Theory* (New York: Verso, 1989).

Stanko, Elizabeth. "The Day to Count: Reflections on a Methodology to Raise Awareness about the Impact of Domestic Violence in the UK," *Criminology and Criminal Justice* 1(2001): 215–26.

Stepan, Nancy L. *"The Hour of Eugenics": Race, Gender, and Nation in Latin America* (Ithaca, NY: Cornell University Press, 1991).

Svendsen, Lars. *A Philosophy of Evil* (Champaign, IL: Dalkey Archive Press, 2010).

Taylor, Kathleen. *Cruelty: Human Evil and the Human Brain* (Oxford: Oxford University Press, 2009).

Taylor, Peter J. "God Invented War to Teach Americans Geography," *Political Geography* 23(2004): 487–92.

Thion, Serge. "The Cambodian Idea of Revolution," in *Revolution and Its Aftermath in Kampuchea: Eight Essays*, edited by David P. Chandler and Ben Kiernan (New Haven, CT: Yale University Southeast Asian Studies, 1983), 10–33.

Thurschwell, Adam. "Ethical Exception: Capital Punishment in the Figure of Sovereignty," in *States of Violence: War, Capital Punishment, and Letting Die*, edited by Austin Sarat and Jennifer L. Culbert (Cambridge: Cambridge University Press, 2009), 270–96.

Totten, Samuel, and William S. Parsons (editors). *Century of Genocide: Critical Essays and Eyewitness Accounts*, 3rd edition (New York: Routledge, 2009).

Tucker, William H. *The Science and Politics of Racial Research* (Urbana: University of Illinois Press, 1994).

Tyner, James A. *The Killing of Cambodia: Geography, Genocide and the Unmaking of Space* (Aldershot, UK: Ashgate, 2008).

———. "Imagining Genocide: Anti-Geographies and the Erasure of Space in Democratic Kampuchea," *Space and Polity* 13(2009): 9–20.

———. "Toward a Nonkilling Geography: Deconstructing the Spatial Logic of Killing," in *Toward a Nonkilling Paradigm*, edited by Joám Evans Pim (Honolulu, HI: Center for Global Nonkilling, 2009), 169–85.

———. *War, Violence, and Population: Making the Body Count* (New York: Guilford Press, 2009).

———. "Genocide as Reconstruction: The Political Geography of Democratic Kampuchea," in *Reconstructing Conflict: Integrating War and Post-War Geographies*, edited by Scott Kirsch and Colin Flint (Aldershot, UK: Ashgate, 2011), 49–66.

———. *Space, Place, and Violence: Violence and the Embodied Geographies of Race, Sex, and Gender* (New York: Routledge, 2012).

Valentino, Benjamin. *Final Solutions: Mass Killing and Genocide in the Twentieth Century* (Ithaca, NY: Cornell University Press, 2004).

Vickery, Michael. "How Many Died in Pol Pot's Kampuchea?" *Bulletin of Concerned Asian Scholars* 20(1988): 377–85.

Villa, Ouk. "A Bitter Life," in *Children of Cambodia's Killing Fields: Memoirs by Survivors*, compiled by Dith Pran and edited by Kim DePaul (New Haven, CT: Yale University Press, 1997), 115–21.

Vriens, Lennart. "Peace Education: Cooperative Building of a Humane Future," *Pastoral Care in Education* 15(1997): 25–30.

Wachsmann, Nikolaus. "The Dynamics of Destruction: the Development of the Concentration Camps, 1933–1945," in *Concentration Camps in Nazi Germany: The New Histories*, edited by Jane Caplan and Nikolaus Wachsmann (London: Routledge, 2010), 17–43.

Wade, Nicholas. *Before the Dawn: Recovering the Lost History of Our Ancestors* (New York: Penguin Books, 2007).

Waller, James. *Becoming Evil: How Ordinary People Commit Genocide and Mass Killing* (New York: Oxford University Press, 2002).

Wander, Philip. "Introduction," in *Everyday Life in the Modern World*, Henri Lefebvre (New Brunswick, NJ: Transaction Publishers, 2002), vii–xxiii.

Weber, Max. *Economy and Society: An Outline of Interpretive Sociology* (New York: Bedminster Press, 1968).

Whittlesey, Derwent. "Sequent Occupance," *Annals of the Association of American Geographers* 19(1929): 162–65.

Wright, John K. "Terrae Incognitae: The Place of the Imagination in Geography," *Annals of the Association of American Geographers* 37(1947): 1–15.

Yang, Dali L., and Fubing Su. "The Politics of Famine and Reform in Rural China," *China Economic Review* 9(1998): 141–56.

Young, Iris M. *Justice and the Politics of Difference* (Princeton, NJ: Princeton University Press, 1990).

Index

Agamben, Giorgio, 24–25, 44–46, 133, 142, 160–61
Agnew, John, 16
Aitken, Stuart, 47
anatomo-politics, 22
Arendt, Hannah, 14, 154
Auschwitz, 33–34, 70, 103

Bailey, Adrian, 19
bare life, 24, 46
Barker, Philip, 84, 96
Bartky, Sandra, 15–16
Bauman, Zygmunt, 3, 155–56
Becker, Jasper, 91, 95, 97
Belzec, 61, 69
Bergen, Doris, 43, 45, 66–67, 72
Bernburg, 63, 65
Binding, Rudolf, 53–54
biopolitics, 20–22, 25, 39, 82, 161
biopower, 22, 23, 39, 44
body, 21–22, 37–41, 45–46, 54, 95–97, 116–19, 123, 145
Brandenburg, 63
Brandt, Karl, 59–60, 61–62
Browning, Christopher, 10, 43, 66–67
Buchenwald, 71
Burleigh, Michael, 8, 51, 56, 61, 63, 65, 158

Cambodia, 3–4, 21, 36, 111–51; and education, 130–32; and erasure, 112–16; and health, 132–40; and letting die, 140–44; and social control, 116–30
Cartwright, Will, 140–41
Chandler, David, 112, 134
Chang, Jung, 96, 102–3
Chelmno, 69
China, 81–105; and agricultural policies, 89–93; and industrialization, 98–100
Chirot, Daniel, 6, 10
collectives: in Cambodia, 121–26, 141; in China, 93–96
communes: in Cambodia, 122–26; in China, 93–96, 102–3
community, 46–51
concentration camps, 70–73
Crampton, Jeremey, 18
Cresswell, Tim, 4
Culbert, Jennifer, 24
Curtis, Bruce, 19–20

Dachau, 46, 70–71
Dahlman, Carl, 16
Danielsson, Sarah, 38, 42
Darwin, Charles, 38–39, 41
Dean, Mitchell, 20

dehumanization, 12, 101
Delaney, David, 153
Democratic Kampuchea, 111
deportation, 37
Dikötter, Frank, 86–87, 89–90, 98,
    103
discipline, 15, 21–22, 44–45, 83–84, 93,
    95, 116–30, 154
Dranove, David, 2–3, 157, 159
Dreyfus, Hubert, 18
Dubow, Saul, 39
Dy, Khamboly, 127, 138

Ebihara, May, 125
Eichmann, Adolf, 36
Einsatzgruppen, 67–69, 71
Elden, Stuart, 16, 18
Em, Sokhym, 134, 143
Esposito, Roberto, 46, 51, 54–55, 73
essence trap, 11
eugenics, 37–41, 50; negative, 40;
    positive, 40
euthanasia, 1, 37, 40, 44, 55, 58–60,
    70, 73, 158–59; and infants, 59–60;
    and Aktion T-4 Program, 60–66, 69,
    72
every day, 2, 153–55, 156, 161
evolution, 38–39

famine, 86, 87–100, 101–3, 104–5
Fein, Helen, 10
Final Solution, 66–73
Flossenburg, 71
Foucault, Michel, 16–17, 20–22, 23–25,
    37, 83–84, 95, 123, 142, 160
France, 132–33
free will, 14–15, 84
Friedlander, Henry, 55, 69
Fritzsche, Peter, 36

Gamson, William, 10
gas chambers, 62, 69–73
Gemeinschaft, 46–47
General Plan East, 66, 71–72
genocide, as modern construct, 155–56

geographical imagination, 5, 6, 7,
    11–12, 25, 41, 130; and Cambodia,
    120–21, 131–32, 144–45; and China,
    83, 88, 100; and Germany, 35–36,
    45, 55, 73
geographies of experience, 153–54
geography, 5, 12, 41, 47, 114, 130–32,
    161
geopolitics, 41–44, 51
Germany, 33–79
Gesellschaft, 46–47
ghettos, 71–73
Gilligan, James, 6, 13
Glover, Jonathan, 158–60
Gnadentod, 61
Goebbels, Joseph, 36
governance, 20
Grafeneck, 63, 65
Great Famine, 36, 86–87; as genocide,
    82–83, 104–5
Great Leap Forward, 10, 81, 87–100,
    113
Green, O. H., 104, 142
Gregory, Derek, 6, 100
Grossman, Dave, 7–8

Hadamar, 63
Halliday, Jon, 96, 102
Hamburg, David, 3
Harris, John, 157–60
Hartheim Castle, 1, 63
Heydrich, Reinhard, 36, 37, 66
Highmore, Ben, 153
Himmler, Heinrich, 35, 36, 37, 45,
    66–67
Hinton, Alexander, 6
Hitler, Adolf, 36, 37, 43, 45, 51, 55,
    59–60, 61, 66, 85, 88, 112
Hobbes, Thomas, 17, 23
Hoche, Alfred, 53–54
Holocaust, 1–3, 8–10, 12, 14, 33–79,
    83, 113, 144, 155
homo sacer, 24, 45, 55, 138
Hope, Tony, 158–60
Horowitz, Irving, 24, 25, 156, 158–59

Ieng Sary, 113
Ieng Thirith, 113, 143
impunity, 100–104
Inwood, Joshua, 153

Jackson, Karl, 144
Jackson, Robert, 16–17
Jewish policy and Nazi state, 66–73
Jost, Adolf, 53, 59
justice, 13, 14, 46
just-world phenomenon, 101

Kahn, Paul, 25, 97
Khieu Samphan, 113
Kiernan, Ben, 3
killing, 6–13, 22, 67–73, 82, 97, 119,
    158–60; geography of, 7–8; justified,
    6, 12, 14, 37, 44, 46, 53, 55, 60,
    105, 127–28; state sovereignty and,
    22–25, 51, 53, 97
Kjellén, Rudolf, 42, 51
knowledge, 37, 47–48; and violence,
    83–87, 102, 105
Koonz, Claudia, 35, 47
Kuper, Leo, 6, 156

Lamarck, Jean Baptiste, 39
Lebensraum, 42–43, 52, 55, 66, 70, 73,
    85, 89, 114
Lebenswert, 53
Lefebvre, Henri, 4, 114, 154
Lemke, Thomas, 18
Lerner, Melvin, 101
"letting die," 23–24, 60, 70, 72, 82–83,
    104–5, 113, 140–44, 156, 157–59
Levi, Primo, 2, 14
LeVine, Peg, 119
life, 1, 4, 13, 22, 133, 161; calculation
    of, 2–3, 18, 33–35, 36, 43, 51, 53,
    58, 61, 73, 88, 99, 123, 138, 140,
    157; disallowal of, 23–25, 44,
    101–3, 104–5, 142, 157–60, 161;
    management of, 1, 2, 15, 24, 86, 89,
    97, 119, 141; unworthy of living,

36, 39, 50–51, 53–55, 56, 59, 60, 62,
    142, 156, 158
Lifton, Robert, 56, 59, 158
Li Fuchun, 88, 104
Liu Shaoqi, 89, 93, 104
Locard, Henri, 116, 126, 142–43
Lodz, 7172

Majdanek, 70
Mam, Kalyanee, 125
Mam, Teeda Butt, 123
Mao Zedong, 81, 83, 85–87, 88–100,
    112–13, 115–16, 120, 145
marriage, 117–20, 122
Mauthausen, 71
McCauley, Clark, 6, 10
McKale, Donald, 73
McMahan, Jeff, 44, 160
morality, 11–13, 157, 161; moral
    exclusion, 11, 47, 53, 100–101;
    moral inclusion, 11, 47

Nazi laws, 46–51
Ngor, Haing, 127, 144
normalization, 21, 37, 119
Noun Chea, 113

Opotow, Susan, 11, 13
Ordover, Nancy, 41, 52
Ó Tuathail, Gearóid, 41
Oveson, Jan, 132, 143

Painter, Joe, 15, 94, 153
Patterson, Orlando, 45–46
Peng Dehuai, 86–87
Pink, Thomas, 15
Pol Pot, 113, 116, 120, 135, 144–45
population(s), 18–22, 24, 35, 37–41,
    73, 145, 157; classifications, 38–39,
    42–43, 47–51, 88, 116–17, 124–25,
    133, 138
principle of non-vicious choice, 157
Proctor, Robert, 57, 63, 158

Quinn, Kenneth, 123

Rabinow, Paul, 18
race, 36, 37–41, 51, 66, 88, 156;
    classifications, 38
Rachels, James, 105, 141–42, 160
Ratzel, Friedrich, 42, 51, 53
Rees, Laurence, 66–67
refrain, 142–44
Rhodes, Richard, 67
Rose, Nikolas, 24, 44, 160–61
Rousseau, Jean-Jacques, 23
Rwanda, 11, 14

Sachsenhausen, 71
Sam, Roeun, 3–4, 124, 127
Sarat, Austin, 24
Schulten, Susan, 130
security centers, 127–28
Semelin, Jacques, 12
Sen, Amartya, 47
sequent occupance, 114
Shapiro, Judith, 85–86
Shaw, Martin, 3, 6, 155
Sibley, David, 47
slippery slope, 158–59
Sobibor, 61, 69
social categorization, 11–12, 101, 117
social death, 45–46
Soja, Ed, 5
Sonnenstein, 63
sovereignty, 4, 14–22, 37, 44, 59; and
    violence, 22–25, 53–56, 97, 140,
    156, 161
space, 4–5, 114–15, 154; will to, 6–13,
    25, 66–73, 88, 114–15, 156, 159
spatiality, 4, 5, 15, 36, 153
Spencer, Herbert, 38, 41
Stanko, Elizabeth, 155
starvation, 70–71, 81–82, 111
state(s), 15–22, 94; and right to take
    life, 22–25, 59–60, 140–44, 145,
    156–57, 161; Cambodia, 111–51,
    156; China, 81–110, 156; Nazi,
    33–79, 155–56
Stepan, Nancy, 39

sterilization, 1, 40, 48, 50, 52–56,
    56–58, 70, 156
surveillance, 21, 102, 123
Svendsen, Lars, 6, 14

Taylor, Kathleen, 11
Taylor, Peter, 130
terra incognita, 5
territory, 16–19, 41–42
terrorist society, 154, 156
thanatopolitics, 24
Thurschwell, Adam, 22–23
torture, 127–28
Trankell, Ing-Britt, 132, 143
Treblinka, 61, 69
Tuol Sleng (S-21), 127–29, 131

Union of Soviet Socialist Republics,
    89–90, 104

Valentino, Benjamin, 82, 94
Vietnam, 112–13
violence, 6–13, 140, 145, 154–55; and
    Cambodia, 126–28; and Maoist
    China, 83–87, 97; and Nazi State,
    51–56; direct, 13, 139; spaces of,
    14–22; structural, 13, 87, 100, 139,
    141
Volk, Volkgemeinschaft, 46–51, 59, 62,
    85

Waller, James, 11–12, 101
Wander, Philip, 153–54
Wannsee Conference, 67
Weismann, August, 39
Whittlesey, Derwent, 114
Wippermann, Wolfgang, 51, 61
Wirth, Christian, 61–62
Wright, John K., 5

Xin Meng, 93

Zhang Wentian, 86–87
Zhou Enlai, 83, 89, 93, 104

About the Author

James A. Tyner received his PhD from the University of Southern California and is professor of geography at Kent State University. His research interests include political, population, and social geography. He is the author of twelve books.